Applied Chaos and Complexity Theory in Education

Şefika Şule Erçetin
Hacettepe University, Turkey

A volume in the Advances in Educational Marketing, Administration, and Leadership (AEMAL) Book Series

Information Science REFERENCE
An Imprint of IGI Global

Published in the United States of America by
Information Science Reference (an imprint of IGI Global)
701 E. Chocolate Avenue
Hershey PA 17033
Tel: 717-533-8845
Fax: 717-533-8661
E-mail: cust@igi-global.com
Web site: http://www.igi-global.com

Library of Congress Cataloging-in-Publication Data

Names: Ercetin, Sefika Sule, editor.
Title: Applied chaos and complexity theory in education / Sefika Sule Ercetin, editor.
Description: Hershey, PA : Information Science Reference, [2016] | Includes bibliographical references and index.
Identifiers: LCCN 2016010968| ISBN 9781522504603 (hardcover) | ISBN 9781522504610 (ebook)
Subjects: LCSH: School management and organization--Mathematical models. | Chaotic behavior in systems. | Organizational behavior.
Classification: LCC LB2806 .A612 2016 | DDC 371.2--dc23 LC record available at https://lccn.loc.gov/2016010968

This book is published in the IGI Global book series Advances in Educational Marketing, Administration, and Leadership (AEMAL) (ISSN: 2326-9022; eISSN: 2326-9030)

British Cataloguing in Publication Data
A Cataloguing in Publication record for this book is available from the British Library.

Advances in Educational Marketing, Administration, and Leadership (AEMAL) Book Series

ISSN: 2326-9022
EISSN: 2326-9030

Mission

With more educational institutions entering into public, higher, and professional education, the educational environment has grown increasingly competitive. With this increase in competitiveness has come the need for a greater focus on leadership within the institutions, on administrative handling of educational matters, and on the marketing of the services offered.

The **Advances in Educational Marketing, Administration, & Leadership (AEMAL) Book Series** strives to provide publications that address all these areas and present trending, current research to assist professionals, administrators, and others involved in the education sector in making their decisions.

Coverage

- Academic Administration
- Advertising and Promotion of Academic Programs and Institutions
- Educational Finance
- Direct marketing of educational programs
- Governance in P-12 and Higher Education
- Enrollment Management
- Technologies and Educational Marketing
- Consumer Behavior
- Academic Pricing
- Educational Management

IGI Global is currently accepting manuscripts for publication within this series. To submit a proposal for a volume in this series, please contact our Acquisition Editors at Acquisitions@igi-global.com or visit: http://www.igi-global.com/publish/.

The Advances in Educational Marketing, Administration, and Leadership (AEMAL) Book Series (ISSN 2326-9022) is published by IGI Global, 701 E. Chocolate Avenue, Hershey, PA 17033-1240, USA, www.igi-global.com. This series is composed of titles available for purchase individually; each title is edited to be contextually exclusive from any other title within the series. For pricing and ordering information please visit http://www.igi-global.com/book-series/advances-educational-marketing-administration-leadership/73677. Postmaster: Send all address changes to above address.

Titles in this Series

For a list of additional titles in this series, please visit: www.igi-global.com

Promoting Global Peace and Civic Engagement through Education
Kshama Pandey (Dayalbagh Educational Institute, India) and Pratibha Upadhyay (University of Allahabad, India)
Information Science Reference • copyright 2016 • 464pp • H/C (ISBN: 9781522500780) • US $185.00 (our price)

Handbook of Research on Organizational Justice and Culture in Higher Education Institutions
Nwachukwu Prince Ololube (Ignatius Ajuru University of Education, Nigeria)
Information Science Reference • copyright 2016 • 571pp • H/C (ISBN: 9781466698505) • US $300.00 (our price)

Developing Effective Educational Experiences through Learning Analytics
Mark Anderson (Edge Hill University, UK) and Collette Gavan (Edge Hill University, UK)
Information Science Reference • copyright 2016 • 361pp • H/C (ISBN: 9781466699830) • US $175.00 (our price)

Handbook of Research on Quality Assurance and Value Management in Higher Education
Walter Nuninger (Université de Lille, France) and Jean-Marie Châtelet (Université de Lille, France)
Information Science Reference • copyright 2016 • 458pp • H/C (ISBN: 9781522500247) • US $295.00 (our price)

Campus Support Services, Programs, and Policies for International Students
Krishna Bista (University of Louisiana at Monroe, USA) and Charlotte Foster (Missouri Western State University, USA)
Information Science Reference • copyright 2016 • 354pp • H/C (ISBN: 9781466697522) • US $185.00 (our price)

Facilitating Higher Education Growth through Fundraising and Philanthropy
Henry C. Alphin Jr. (Drexel University, USA) Jennie Lavine (Higher Colleges of Technology, United Arab Emirates) Stormy Stark (Pennsylvania State University, USA) and Adam Hocker (Pennsylvania State University, USA)
Information Science Reference • copyright 2016 • 372pp • H/C (ISBN: 9781466696648) • US $195.00 (our price)

IGI GLOBAL
DISSEMINATOR OF KNOWLEDGE
www.igi-global.com

701 E. Chocolate Ave., Hershey, PA 17033
Order online at www.igi-global.com or call 717-533-8845 x100
To place a standing order for titles released in this series,
contact: cust@igi-global.com
Mon-Fri 8:00 am - 5:00 pm (est) or fax 24 hours a day 717-533-8661

To Nihan and Şuay Nilhan with endless love.

Editorial Advisory Board

Table of Contents

Detailed Table of Contents

It is mainly concentrated on variables which prevent girls' education in this study. The aim of this study is to describe chaos, complication and antidemocratic practices that emerge on girls' education. A wide comprehensive literature survey is performed on variables which concern the variables of girls' education for the scope of the study. This study is a survey designed model that aims to define the properties concerning the girls' education. Many social, economic, political and cultural problems create negative effects on girls' education. Not being able to make democracy a life style brings along girls' not being able to make use of education as it should be which is a human right. As a result, it is emphasized that only "whole democratic life style" practice for all could solve the problem of "social inequality of girls" which is enduring for ages.

In this study, an attempt was made to explore and examine the influence of changing paradigms on educational management. In approaching the chapter, the authors discussed the status-quo by describing the traditional management paradigm. The drivers to change which triggered off the rise of a new paradigm were highlighted and these dully led to a description of the new paradigm whose principles and theories were fully examined. The chapter then engaged a detailed analysis of the various changes in the education management and school administration field consequent to the shift in paradigms. The authors feel that an understanding of the major paradigms in education as well as those critical to educational management and administration is a universal calling. There is no gain saying, therefore, that this chapter will go a long way into improving educational management and administration both at structural level and at the school level.

Increasing environmental complexity and ambiguity require organizations to rely on their members' proactive behaviors to deal with potential chaotic occurences surrounding the organizational functioning. Individuals respond these occurences by displaying various proactive behavior forms to build predictability islands in ocean of uncertainty. In this context, principal proactivity is one of key determinants of effectiveness of schools functioning in complex and anarchic environments. In this study, numerous forms of proactive behaviour were categorized under six general dimensions as innovativeness, issue selling, social networking, strategic scanning, strategic learning and feedback seeking. A causal comparative design was used to determine levels of general forms of proactive behavior performed by school principals. The data were collected from school principals working at different levels of public and private schools in İstanbul Province during 2014-2015 year. According to the results of study, the most frequent form of proactive behaviour displayed by school principals was strategic learning.

This study examines that in what aspects the levels of leaders' emotional intelligence in chaos and complexity situations can affect the organizations. Also, in this study the dimensions of emotional intelligence which are closely related to social life and organizational systems will be considered in terms of chaos and complexity theories. At the same time in this study, how a situation that appears to be complex but has an order inside and has the potential to be directed with small interventions can be handled by a leader that has a high capacity of using his/her emotional intelligence will be scrutinised. Additionally, this study will present the advantages of some leader capabilities (such as personal and interpersonal relationships, stress management, adaptability, general mood, empathy, motivation and self-awareness) in the management of the organizations in the context of chaos and complexity theories. Last but not least, the effect of emotional intelligence level in the rising of new leaders will be mentioned in this study.

The university system is an organized anarchy. It is a system that brings different components from different dimensions with different focus sometimes conflicting one another in one place, in order to achieve a particular goal. This paper examines the university as a system that comprises different faculties, departments and units managing different disciplines and courses/subjects. Chaos and complexity nature of university was examined and results show that each faculty and department is taking a different direction without caring about others, and in the same faculty there are different departments handling different courses. And lecturers are using different methods in the conduct of their lessons and researches, and all these struggles are made towards achieving university objectives. The challenges faced by university system and possible solutions to overcome those challenges were discussed.

Stigma takes place as a concept in the literature that affects individuals' lives negatively as well as can disrupt chemistry of organization and inhibit organizational behavior. Stigma is a concept that ostracizing individuals or groups because of their features that are not accepted by society and restricted their life on all areas. Studies on stigma have been increased during twenty years. Although school principals play the lock role and the most affecting actor in positively or negatively to school, studies on principals' stigma tendency are extremely limited. Stigma tendencies of principals' can affect chemistry of organization much more than stigma tendencies of teachers. The purpose of the study is to determine stigma tendencies of principals' by using quantitative research methods. According to research results; principals' stigma tendencies are significantly different according to seniority, level of religious feelings and health status.

The interest to all actors and observers within educational arena denotes a long-term trend of expansion of educational professionalism accompanied by a continuous debate about its desirability, and a perennial instability or dynamic of the structures within education systems. Over the years, however, the emphasis placed on shape of professionalism in education systems has varied substantially to reflect the complexity contained therein from a purely managerial and leadership perspective. Moreover, we observe considerable changes of ''views and controversies'' concerning the most desirable quality and structure of teacher interaction and professionalism in educational institutions. This chapter explores the state of teacher interaction and collaboration in educational institutions as well as dissecting implications of same for management and leadership in schools. The study is deemed an eye opener to managers, leaders in schools to 'mind the gap' that is teacher insularity if they are to effectively achieve the very best out of them.

Chaos and complexity theory has been used in the study of the natural sciences for over thirty years. Throughout the years, experts from various fields have used it as a new way to view the world around them, including its applications to the field of education and subsystems within the education system. However very few studies have been conducted on the application of chaos theory to classroom discipline. The field of classroom discipline, like the natural world, can also be observed from this perspective because it exhibits many features of chaotic/complex systems. Classrooms are often described as complex, dynamic and unpredictable environments. This makes it difficult for teachers to understand and manage classroom discipline. This paper explores the complexity of classroom discipline and how the principles of chaos and complexity theory reflect on classroom discipline. This would help inform management strategies and alternatives that would in turn enhance and improve student academic achievement and overall performance.

Corruption has caused untold havoc not only to the quality, quantity of education and credibility of citizens in Nigeria, but also to our national growth and development in all aspects of human endeavor. In academia it diffuses and circulates through the aorta and capillaries of the system, such that it is no longer view as a bottleneck thwarting the successful implementation of educational policies. The ugly scenario has degraded our educational foundation popularly called lower basic created a permanent vacuum which the subsequent higher level proved helpless to fill. The objectivity, reliability, trust, transparency, sincerity and above all integrity known in the academia as its "glory" have been wiped out by corrupt practices. At this junction, it is mandatory to dig out the reasons behind indulgence in corruption in our lower basic, identify its causes, approaches and effects as they are dimensional in nature. The paper will attempt the definitions of the term corruption at initial, and the concluding aspect will suggest the possible remedies.

Examinations occupy a strategic position in our lives and society today. In a situation where moral principles, rules and regulations for conducting examinations are truncated by either teacher or the learner, the validity and reliability of examination and certification are at risk. The nature and process of examination is something that should be looked into deeply having known the role of examination in our lives. That is essence of this paper; the way high-stakes examinations have affected various aspects of the education system in the realm of teaching, learning, assessment, discipline, roles of stakeholders, teachers' beliefs, community expectations, functional role of education, educational outcomes, individual differences, psychological status of learners, educational returns, school completion etc. This paper articulated these issues in terms of using high-stakes examinations as a measure of accountability in the education system and premised on the level of faith placed on these high-stakes examinations irrespective of educational value thereof.

Conflicts are inevitable and very common in educational institutions the world over. Conflict happens whenever and wherever in HEIs. Conflicts also rampantly occur at HEIs. Conflicts are therefore inevitable although this should not cause alarm truth is that they are natural and have benefits to organizational progress. It can therefore be noted that conflict management is part and parcel of school administration challenges that HEIs managers should handle as part of institutional politics, with extra care to benefit system and its workers. Therefore, experienced managers should motivate different conflicting groups to improve their methods and attitude towards work so that organizational goal can be achieved in the most effective manner. This however is a tough undertaking since it requires technical knowledge and skills on the part of leader. Accordingly, understanding complexity theory and applying it effectively in organizational management can transform conflict into an effective catalyst for improving school system.

Collegial groups of professional people have been shown to produce high adult achievement and performance and therefore learning the skills of working with groups to solve instructional problems is a critical task of supervision. This paper covers knowledge, skills, and procedures for developing productive instructional improvement groups in the complex school environment characterized by turbulence and chaos emanating from dysfunctional group members. Throughout this paper, the importance of these methods in developing group leadership skills has been explained: using group observations, changing group leadership styles, dealing with dysfunctional members, and preparing for meetings. The paper also demonstrates how development supervision can be applied to professional groups by showing how a developmental supervisor matches the supervisory approaches to the group's characteristics.

As it is stated that the economy of a country can never be greater than the education of its people, today parents and several organizations are investing a lot of money in education to better the future of their children and also meet the demands of the several generations to come. Through this process, the level of competition among higher education institutions is too high for them to survive and continuously attract parents, citizens and funders to meet demands of both institutional and community challenges. The aim of this paper, therefore, is to explore the concept of decision making, factors that determine the decision making process, the effectiveness of decision making and problem solving in higher education institutions deemed to be largely chaotic and complex. World over, government and states are investing in higher education to transform their communities in terms of knowledge, skills, culture and economic development.

This study is designed to provide a picture of principal leaders in the school. The focus of this study is to identify and describe the characteristics of principal's leadership process in complex situation based on self-assessment. The study conducted by qualitative approach consisted of volunteer principals working in the city center of Ankara, Turkey. We discuss how these findings provide a deeper understanding of the latest principals' leadership process in complexity based on principals' self-assessment. The researcher gathered and analyzed data in an effort to create a template of essential leadership characteristics that could be utilized at other schools or in other educational environment. The results likely to suggest that self-assessment of principals in complexity are well worth revealing how the principals perceive their leadership process and what leading characteristics are viewed when they have complex situation.

Foreword

What do the sciences of nature have to offer for social sciences? This question has been answered by thousands of different theories throughout the study of the social interactions of the mankind. To create an order out of chaos has been the gist of history from an Orwellian perspective. Yet, the times of chaos and wars have usually resulted in long-lasting orders whether or not the chaotic environment is a consequence of an *Ordo Ab Chao* conspiracy. Chaotic situations and effects of them, of course, have not initiated with the Chaos and Complexity Theory. Coping with chaotic and complex circumstances and explaining them have been the leading endeavors fueling the scientific achievements. However, with the information age, the speed of dispersion of any chaotic and complex situation encountered at a given place of the world, has a plethora of repercussions, creating a network of dynamic interactions, often increasing the entropy of the global social processes. Therefore, understanding the humanities and the social relations, the communities and the societies and their complexities has become a Sisyphean challenge for a social scientist. While a deterministic approach could be the key to a theory, the social sciences had to learn to cope with unpredictability and the chaotic nature of endless human interactions. If we take the famous question of Edward Lorenz, "Does the Flap of a Butterfly's Wings in Brazil set off a Tornado in Texas?" and ask with regard to, say, political sciences, "Does a protest clap of two hands in a country set off a Revolution in another country in far continent?" What would be the methodology to answer it. If we are not going to suffice to give a response through prejudicial theoretical explanations restricted to our cognitive ability based on our limited historical and social accumulation of knowledge on the conditions that interweaves the two phenomena, then, even to be able to find a trivial answer for this question, the amount of information necessary to be collected would only be calculated through demanding complexity computations.

Chaos and Complexity Theory, today, appear as a possible code that may help to unlock the mysteries of the nature of the social and political interactions of the mankind. 'Chaos' and Chaos Theory, a field of study in Mathematics and in other disciplines such as Physics, Engineering, Economics, Management, Education and other fields may become the keyword to define the social and political sets of circumstances dependent on various set of parameters and determinants, and therefore started to be applied in social sciences recently.

Internationally prominent scholar, *Şefika Şule Erçetin*, has edited an awe-inspiring book dealing with the roots and consequences of our dynamically transformed social, political, educational and economic life, bringing together a myriad of thought-provoking essays over very interesting themes. Her eminent contribution to academic world assures a valuable scholarship in the following chapter.

Applied Chaos and Complexity Theory in Education includes 16 chapters on Education.

This book opens new insights in education sciences. In the first chapter *Mehmet Özbaş* analyzes the nature of the problems in chaotic and complex situations in girls' education and offers some solutions to the "social inequality of girls". In second chapter *Şefika Şule Erçetin and Ssali Muhammadi Bisaso* dwell on the influence of changing paradigms on educational management and school administration. *Tuncer Fidan and Ali Balcı* help one understand principal proactivity genuinely in their insightful chapter. Erçetin, Eriçok and Eranıl's paper is going to present the advantages of some leader capabilities in the management of the organizations in the context of chaos and complexity theories. *Yusuf Abubakar and Abubakar Mansur Danjeka from Nigeria* explain the university system as a mere organized conflict environment. *Then, Halime Güngör investigates* principals' views on stigma tendency *and* consequences of the stigmatization process in her chapter. *Sa'adu Isa Bashar and Zayyanu Sambo* elaborates onkey issues with regard to managing the chaos and complexities of informal organizations for the effectiveness of schools as formal organizations in the next chapter. *Kenan İli and Mehmet Menteşe* invite us to dwell deeply on education and leadership in Social Networks. *Ssali Muhammadi Bisaso* explains the methods to comprehend complexity of school leadership in the wake of teacher insularity. *Mehmet Menteşe and Kenan İli* find answers on how to manage or inspire your perception about social media and the perception management as a power struggle. Furthermore, *Ogunlade Joseph Olurotimi* makes a critical review of the complexities of breastfeeding and its attending effects on the health of the mother and child. *Fathimath Saeed* explores the complexity of classroom discipline and how the principles of chaos and complexity theory reflect on classroom discipline in her chapter *Mukhtar Salihu Nawait* attempts the definitions of the term corruption at initial and the concluding aspect will suggest the possible remedies on quality and quantity of education in Sokoto State, Nigeria. *Ssali Muhammadi Bisaso and Shakira Bodio* articulate complexity ramifications in terms of using high-stakes examinations as a measure of accountability in the education system and premised on the level of faith placed on these high-stakes examinations irrespective of educational value thereof. *Farooq Miiro* profoundly explains conflict management in higher educational institutions with complexity perspective. *Muhamadi Kaweesi from Uganda* comprehends leadership skills in complex collegial adult groups and explains knowledge, skills, and procedures for developing productive instructional improvement groups in the complex school environment characterized by turbulence and chaos emanating from dysfunctional group members. *Muhamadi Kaweesi and Farooq Miiro* explore the concept of decision-making, factors that determine the decision-making processes, the effectiveness of decision- making and problem solving in higher education institutions which are deemed to be largely chaotic and complex. Last chapter of education section written by *Leyla Yılmaz Fındık* focuses of this study is to identify and describe the characteristics of principal's leadership process in complex situation based on self-assessment.

The latest developments in technology and communication, existing industrial problems and new socio-economic expectations have led to the emergence of more chaotic and complex conditions globally. This collection of significant essays outlines the circumstances in various countries in different contexts and marks a key contribution to the application and understanding of chaos and complexity in social sciences. Being an original contribution to the conceptualization of politics, economics, organization and management, education and literature through different lenses, this unique work enhances the role of *Applied Chaos and Complexity Theory in Education*. Offering many novel ideas and reflections for the future, this volume aims to be inspirational for many scholars in their attempts to explore new tools and means for scientific thought.

I would like to sincerely express my gratitude to the contributors as well as the editor, Şefika Şule Erçetin for the compilation of such an extensive volume which, I presume, will be a reference book for the future of interdisciplinary studies on the education.

Ege Erkoçak
Ministry for EU Affairs, Turkey
Ankara, January 2016

Ege Erkoçak *is the Director for Political Affairs of the Ministry for EU Affairs, Turkey. He holds an M.A degree with his thesis titled "An Ever Enlarging Europe: Enlargement of the EU, 1990s and Turkey." He held several seminars on "EU Enlargement and EU Negotiations" and "Judiciary and Fundamental Rights" and "Justice, Freedom and Security" issues delivered in Bilkent University, Middle East Technical University, Ankara University, Bosphorus University, Bahçeşehir University, Bilgi University. He has written articles on development of çivil society, meta-governance and EU enlargement.*

Preface

There has been a fluent daily routine of life for people, societies, countries and also for the world. These individuals, groups and societies have been lived chaotic, risky, crisis situations on this daily routine in short-term or long term. Chaotic situations and effects of them of course has not begun with the chaos and complexity Theory, chaos in life of individuals has been lived since the first human being in the world but solutions and coping strategies with these chaotic and complex situations has been changing with scientific developments. However, as a result of digital revolution, the average profile of world has changed as well. The repercussions of an event anywhere in the world has been spreading so quickly to all over the world, so coping with this transformation it is important to have and requires an effective managing power to determine the fate of the country.

In this sense, responsibility and mission of scientists to human and future of the world being is support managing power of the world and human with scientific developments that useful for human. So, 'chaos' and chaos theory, though it is a field of study in mathematics and in other disciplines such as physics, engineering, economics, management, education and other fields has been the key word to define the complex alterations and situations, and applied in social sciences recently.

All chapters in this book contribute chaos and complexity theory in different perspectives and also solution of the universal problems. There are sixteen chapters in the book. Chapters are about chaos and complexity in education.

This book combines so many researches from different disciplines in all over the world. Therefore you will not only find chaos and complexity theory in terms of phenomena on that create difference in scientific but also usable and useful knowledge to understand social, natural reality and also reflections for future in terms of chaos theory in this unique book. We believe that readers would enjoy studies related to chaos, complexity.

Şefika Şule Erçetin
Hacettepe University, Turkey
January 2016

Chapter 1
Chaotic and Complex Situations in Girls' Education:
Problems and Solutions

Mehmet Özbaş
Erzincan University, Turkey

ABSTRACT

It is mainly concentrated on variables which prevent girls' education in this study. The aim of this study is to describe chaos, complication and antidemocratic practices that emerge on girls' education. A wide comprehensive literature survey is performed on variables which concern the variables of girls' education for the scope of the study. This study is a survey designed model that aims to define the properties concerning the girls' education. Many social, economic, political and cultural problems create negative effects on girls' education. Not being able to make democracy a life style brings along girls' not being able to make use of education as it should be which is a human right. As a result, it is emphasized that only "whole democratic life style" practice for all could solve the problem of "social inequality of girls" which is enduring for ages.

INTRODUCTION

In this section of the study, reasons of the chaos and complexity experienced for the education of female children were emphasized. In this sense, familial, social, cultural, political, and administrative variables that prevent female children from education especially in democratic management systems were discussed.

DOI: 10.4018/978-1-5225-0460-3.ch001

Women and girls have always been the social side of humankind that have been abused and neglected throughout the history of humanity. This claim cannot be considered as exaggerated. Even in democratic countries, humankind entered the New Millennium (3rd Millennium) without finding a solution to the problem of equality of women and men (Buyruk, 2008; Gage, 2013; Kocacık & Gökkaya, 2005; Taner & Gökler, 2004). The chaos and complexity female children experience in terms of social, political, economic, and cultural aspects is one of the basic problem areas of democratic management systems. Democratic regimes have to find multiple and pluralistic solutions to this "chronic social problem" through a "chaotic management approach" (Erçetin & Maya, 2005; Ereş, 2006; UNESCO, 2010). Female children's not benefiting from educational opportunities properly is the leading factor on the basis of problems they experience.

It is necessary to look for the reasons that prevent education of female students in direct or indirect effects of socio-economic variables families have. Female children are negatively affected from administrative implementations related to education such as the reasons arisen from familial and social surrounding properties as the leading. This causes chaos and complexity in their education. The situations that create chaos in education of female children are: Socio-economic situation of families, negative social surrounding variables, inadequacy of the income level, and low level social status and education of parents, especially of mother (Wallace-Bruce, 2010). Moreover, inefficiency of schools, especially pre-school teaching and compulsory teaching, is another important variable (Özbaş, 2009; Özbaş, 2010).

Purpose of the Study

Management systems are obliged to actualize the implementations that will reveal the satisfaction level and quality of services they render for their societies (Taş & Yenilmez, 2008). In this sense, democratic management systems have important administrative obligations. When considered in terms of social aspects, it is clear that there are significant discrimination activities in implementations related to female gender. One of these implementations is the chaos, complexity and problematic situations encountered by female children in education. Female children cannot equally benefit from educational opportunities properly according to the social classes, and they experience important problems. These problems cause chaotic situations in time. In terms of democratic management implementations based upon social justice, it is necessary to discuss in comprehensive studies to what extent female children benefit from democratic educational opportunities. For that reason, it was aimed to emphasize chaotic situations encountered in education of female children in this study.

METHODOLOGY

In this study, the situations that prevented female children from education were emphasized. Within the scope of the study, a comprehensive literature review was primarily carried out upon the variables that affect the education of female children. For that reason, this research was a study on screening model that described the current situation related to the educational properties of female children. The researches in screening model are the ones including making situational generalizations and descriptions through collecting data related to what the results and findings revealed on any subject or subjects are (Balcı, 2010; Cohen, Manion & Morrison, 2007; Karadağ, 2010; Karasar, 2010). During the reference review process, the situations related to the education of female children were designed considering the universal democratic criteria. These comprehensive international studies were primarily emphasized.

Gender Discrimination to Female Children

Gender discrimination for girls and women has been one of the most important problems that humankind has not found a solution for ages. The differentiation and inequalities among the social classes, cultural and ideological paradigms, socio-economic status, and social, economic, political and legal insufficiency cause this gender discrimination for girls (Coleman & McNeese, 2009; Özbaş, 2012).

In democratic countries, total social development can be created through providing everyone to benefit from fundamental rights and freedom without any gender discrimination. In this sense, female children, as well as the males, should be trained as versatile. However, one of the reasons for underdevelopment and not providing total social development democratically is not providing female children to benefit from fundamental human rights and freedom properly. In terms of fundamental rights and freedom, depriving female children from especially their educational rights is one of the most important reasons of underdevelopment. Preventing female children from education through gender discrimination has been one of the most important chaotic situations that negatively affect social development. Negative discrimination for female children is one of the reasons that cause underdevelopment of society (Gleitman, Fridlund & Reisberg, 2000; Kasliwal, 1995; Özbaş, 2012; The World Bank, 2006). In the 21st century, economic, political, and social rights of individuals, their social-economic status, and everything that is related to their freedom and personal development is directly or indirectly based upon education (Brien & Stelmach, 2009). For that reason, everybody should benefit from compulsory educational opportunities, education as the leading, at the highest level.

One of the most important problems of all anti-democratic, underdeveloped or developing countries is their not providing educational opportunities for everyone. One of the reasons that affect this is not considering the education of female children as significant. The basic factor that causes this problem is gender discrimination for female children (Ergün, 2011; Gough & Scott, 2007; Kaynak, 2014; Özbaş, 2014). Not providing women-men equality both socially and legally for ages prevented female children to be in social competition on the bottom of the ladder. This fact negatively affects female children in terms of social opportunities, primarily. Underdevelopment of female children in lower socio-economic classes rather than the ones in higher socio-economic classes is the peak point of social inequality. Conditions of female children that include chaotic properties is a "chronic social disease that has continued since generations," as well.

Socio-Economic Status

The position individuals obtain in reference to their social surrounding properties and their families in their social life can be defined as social surrounding. Pre-condition of a democratic and equal socio-economic status is to be away from any kinds of discrimination. In democratic management systems, socio-economic status of individuals depends upon rights and freedom obtained through education. Political parties as an inevitable factor of a democratic political life are important to emphasize equality and justice during the process of individuals' obtaining their socio-economic status (Black, 2007; Holmes-Smith, 2006; TEDMEM, 2015; Jones & Coleman, 2005).

Inequalities among the social classes create chaos and affect female children negatively especially in education. Inequality problems of female children based upon gender is generally determined by cultural, ethnical, and socio-economic factors. Female children of several ethnic and social groups all around the world have been noticed to be exposed to gender discrimination. In this sense, especially female children of Roman families can be suggested to be at a very bad situation.

It is essential that, in democratic societies, female that have high socio-economic status and have active roles in areas of life have struggled for their fellows that have no similar rights and opportunities they have. However, this important requirement can be suggested not to be met. Because, rapid and efficient developments have not been experienced in situations of disadvantageous female children as well as in female children that have high socio-economic level. Chaotic situations related to education and participation of female children to the social life has maintained their presence negatively even in a democratic system.

SOLUTIONS AND RECOMMENDATIONS

The chaos, complexity, and problems related to the education of female children are arisen from gender discrimination, negative administrative approaches, and disadvantageous life conditions. In this sense, inefficient, unsuccessful, and non-transparent implementations of especially democratic education and school managements have important effect. In a democratic system, political decision makers and the political structure that has the power of management should actualize accountable implementations. The suggestions below were developed in order to reveal the chaos and complexity related to the education of female children:

- "The principle of superiority of law" should be implemented completely, and fundamental and compulsory educational opportunities should be provided to all female children.
- Democratic education and school managements should declare their schooling rate of female children in fundamental and compulsory education process through a transparent understanding.
- Socio-economic family analyses should be provided for all female children in compulsory educational age population.
- "Student satisfaction researches" upon to what extent female children are benefited from democratic education implementations should be carried out.
- To what extent female children from lower socio-economic classes are represented in social, economic and political life should be compared with female children from higher socio-economic classes, and this should be released to the public.

FUTURE TRENDS

In accordance with the results obtained from this research, the suggestions below were offered for further future researches:

- The obligations democratic education managements are required to fulfill for education of female children should be discussed in comprehensive studies.
- Both national and international public opinion should be created upon the education of female children from disadvantageous and lower-social classes.
- The situations that prevent female children from education should be searched and shared with democratic public opinion in reason-result relationship.

REFERENCES

Balcı, A. (2010). *Sosyal bilimlerde araştırma yöntem teknik ve ilkeler (7. baskı).* Ankara: Pegem Akademi.

Black, R. (2007). *Overcoming disadvantage through the innovative classroom.* Education Foundation Australia. Retrieved from http:www.educationfoundation.org.au

Brien, K. & Stelmach, B. L. (2009). Legal and cultural contexts of parent-teacher interactions: school councils in Canada. *International Journal about Parents in Education, 3*(1), 1-14.

Buyruk, H. (2008). *Eğitimde yaşanan toplumsal eşitsizliklere ilişkin biyografik bir araştırma. Yayımlanmamış Yüksek Lisans Tezi.* Ankara: Ankara Üniversitesi Eğitim Bilimleri Enstitüsü Eğitim Bilimleri Anabilim Dalı Eğitim Ekonomisi ve Planlaması Programı.

Cohen, L., Manion, L., & Morrison, K. (2007). *Research Methods in Education* (6th ed.). New York: Routledge.

Coleman, B., & McNeese, M. N. (2009). From home to school: The relationship among parental involvement, student motivation, and academic achievement. *The International Journal of Learning, 16*(7), 459–470.

Daniel, A., & Moudic, Y. (2010). *Female and gender leadership: To which extents gender impacts on leadership and organizations in France?* (Master dissertation). Linnaeus University.

Erçetin, Ş. Ş., & Maya, İ. (2005). Meslek-aile profilleri ile kadın ilköğretim okul müdürlerinin yönetsel görevler üstlenmelerinde karşılaştıkları engeller. 28-30 Eylül 2005 Denizli-Türkiye. XIV. Ulusal Eğitim Bilimleri Kongresi. Pamukkale Üniversitesi. Türkiye: Denizli.

Ereş, F. (2006). Türkiye'de kadının statüsü ve yansımaları. *Gazi Üniversitesi Endüstriyel Sanatlar Eğitim Fakültesi Dergisi, 19*, 40–52.

Ergün, M. (2011). Eğitim ve kalkınma. Dicle Üniversitesi Sosyal Araştırmalar Merkezi 3. Sosyal Bilimler Sempozyumu: Bölgesel Kalkınmada Eğitimin Rolü, 1-15. Diyarbakır: Sosyal Araştırmalar Merkezi.

Gage, A. J. (2013). Association of child marriage with suicidal thoughts and attempts among adolescent girls in Ethiopia. *The Journal of Adolescent Health, 52*(5), 654–656. doi:10.1016/j.jadohealth.2012.12.007 PMID:23433537

Gleitman, H., Fridlund, A. J., & Reisberg, D. (2000). *Social development. Basic Psychology* (pp. 480–517). New York: Norton.

Gough, S., & Scott, W. (2008). *Higher education and sustainable development: Paradox and Possibility*. London: Routledge.

Holmes-Smith, P. (2006). *School socio-economic density and its effect on school performance. In SREAMS School Research Evaluation and Measurement Services. Australian-Sydney*. New South Wales Department of Education and Training.

Ishii, Y. (2003). *Development education in Japan: a comparative analysis of the contexts for its emergence, and its introdiction into the Japanese school system*. New York: Routledge Falmer.

Jones, P. W., & Coleman, D. (2005). *The United Nations and education. Multilateralism, development and globalisation*. New York, London: Routladge Falmer. doi:10.4324/9780203421499

Karadağ, E. (2010). Eğitim bilimleri doktora tezlerinde kullanılan araştırma modelleri: Nitelik düzeyleri ve analitik hata tipleri. *Kuram ve Uygulamada Eğitim Yönetimi*, *16*(1), 49–71.

Karasar, N. (2010). *Bilimsel araştırma yöntemi (21. Baskı)*. Ankara: Nobel Yayın Dağıtım.

Kasliwal, P. (1995). *Development economics*. S. Western Collage Publishing.

Kaynak, M. (2014). Türkiye'de çocuk gelin sorunu. *Nesne*, *2*(3), 27–38.

Kocacık, E. ve Gökkaya, V. (2005). Türkiye'de çalışan kadınlar ve sorunları. *Cumhuriyet Üniversitesi İktisadi ve İdari Bilimler Dergisi*, *6*(1), 195–219.

Özbaş, M. (2009). *İlköğretim okulu yöneticilerinin okul-aile ilişkileri konusunda yapmaları gereken ve yapmakta oldukları işler. In Yayımlanmamış Doktora Tezi*. Ankara: Hacettepe Üniversitesi Sosyal Bilimler Enstitüsü.

Özbaş, M. (2010). İlköğretim okullarında öğrenci devamsızlığının nedenleri – The reasons of absenteeism in primary schools. *Eğitim ve Bilim – Education and Science, 35*(156), 32-44.

Özbaş, M. (2012). Kız çocuklarının ortaöğretimde okullaşma oranlarına etki eden nedenlere ilişkin algıları - Perceptions of girls related to the reasons that affect schooling rates in secondary education. *IJTASE International Journal of New Trends in Arts, Sports & Science Education*, *1*(4), 60–71.

Özbaş, M. (2014). İlköğretim okulu yöneticilerinin aile memnuniyetini karşılama düzeyine ilişkin veli görüşleri. *Eğitim Bilimleri Araştırmaları Dergisi- Uluslararasi E-Dergi-Journal of Educational Sciences Research-International E-Journal. EBAD-JESR, 4*(1), 243-258. 10.12973/jesr.2014.41.12".2014.41.12

Özbaş, M., & Avcı, M. (2013). *Yoksul aile çocuklarının kamusal eğitimden yararlanma düzeylerine ilişkin veli algıları.* IASSR European Conference on Social Science Research. Marmara University, School of Business Administration, İstanbul.

Taner, Y., & Gökler, B. (2004). Çocuk istismar ve ihmali: Psikiyatrik yönleri. *Hacettepe Tıp Dergisi, 35*, 82–86.

Taş, U., & Yenilmez, F. (2008). Türkiye'de eğitimin kalkınma üzerindeki rolü ve eğitim yatırımlarının geri dönüş oranı. *Eskişehir Osmangazi Üniversitesi Sosyal Bilimler Dergisi, 9*(1), 155–186.

TEDMEM. (2015). *Siyasi partilerin 2015 genel seçim bildirgeleri: Eğitim politikalarına ilişkin değerlendirmeler. In Güncel Yayınlar G3.* Ankara: Türk Eğitim Derneği TEDMEM.

The World Bank. (2006). *World Development Report 2007: Development and the Next Generation.* The International Bank for Reconstruction and Development The World Bank.. DOI: 10.1596/978-0-8213-6549-6

UNESCO. (2010). *Why gender equality in basic education in Pakistan? United Nations Educational, Scientific and Cultural Organization.* Islamabad: UNESCO Office, Serena Business Complex.

Wallace-Bruce, N. L. A. (2010). *Assessment of factors affecting female participation in senior high school education in Ghana: A case study of Asunafo North and Sunyani Municipalities and Kumasi metropolitan area. Master of Science the School of Graduate Studies.* Ghana: Kwame Nkrumah University of Science and Technology, in Development Policy and Planning Department of Planning College of Architecture and Planning.

Wang, D. B. (2004). Family background factors and mathematics success: A comparison of Chinese and US students. *Educational Research, 41*, 40–54.

ADDITIONAL READING

Akyüz, E. (2001). Çocuk Hakları Sözleşmesinin temel ilkeleri ışığında çocuğun eğitim hakkı. *Milli Eğitim, 150*, 69–73.

Bener, Ö. (2009). Annelerin kız çocukların eğitimi konusunda görüşleri. *Uluslararası İnsan Bilimleri Dergisi*, *6*(2), 44–59.

Dimbisso, T. S. (2009). Understanding female students' academic performance: an exploration of the situation in south nations nationalities and peoples regional state Ethiopia. Masters of Arts in Development Studies Public Policy and Management (PPM). The Netherlands: The Hague.

Gage, A. J. (2013). Association of child marriage with suicidal thoughts and attempts among adolescent girls in Ethiopia. *The Journal of Adolescent Health*, *52*(5), 654–656. doi:10.1016/j.jadohealth.2012.12.007 PMID:23433537

OECD. (2012). *Equity and quality in education: supporting disadvantaged students and schools. A Global Cities Education Network Repoert*. Asia Society Partnership for Global Learning.

Özbaş, M. (2011). İlköğretim okullarının sosyo-ekonomik açıdan dezavantajlı özelliklere sahip öğrenci ailelerinin eğitim ihtiyaçlarını karşılama düzeyine ilişkin veli algıları. EYFOR II. Eğitim Yönetimi Forumu 2023'e Doğru Eğitim Liderliği. EYFOR II. Bildiriler Kitabi (ss. 60-70), EYUDER Eğitim Yöneticileri ve Uzmanları Derneği, 22 Ekim 2011 Başkent Öğretmenevi, Ankara-Turkey.

Özbaş, M. (2012). Student and parental perceptions on meeting the educational needs of the disadvantaged students in the primary schools. *Educational Research*, *3*(3), 311–319.

Schultz, T. P. (2002). Why governments should invest more to educate girls. *World Development*, *30*(2), 207–225. doi:10.1016/S0305-750X(01)00107-3

UNESCO. (2006). *Getting girls out of work and into school: brief policy*. Bangkok, Thailand: UNESCO Asia and Pacific Regional Bureau for Education.

KEY TERMS AND DEFINITIONS

Abuse of Female Children: This indicates the negative discrimination female children encounters in several aspects such as social gender inequality in underdeveloped, anti-democratic, or developing countries. There are several factors affecting this situation, those are; not providing the superiority of law, socio-economic inequality, social and cultural disadvantages, misinterpretation of religious beliefs, considering female children as secondary beings rather than the male children, etc.

Administrative Discrimination for Female Children: This is not providing female children benefit from private and public opportunities according to social law state rules. Negative discrimination firstly starts at schools in the educational system in sense of female children's having right to use private or public opportunities. Education and school managements could not provide female students benefit from the sources affiliated to education in all grades of education, especially during the grades of compulsory education. Management service cannot be rendered to female children in a way they will improve themselves as versatile in terms of social, personal, democratic, and cultural aspects.

Disadvantageous Female Children: The parents, especially the mother, have lower educational status. Female children of families with temporary, insufficient income and without social security are also included in this sense. The social surrounding opportunities of the area they live are limited, and they have friends who could not internalize the importance of education.

Inequality among the Social Classes: These include the inequalities arisen from the social class they belong to due to socio-economic conditions of families, their social life surrounding, and their political, ethnical, etc. properties. Individuals find themselves in a social class from birth without having the opportunity of selection. The social class that is especially disadvantageous due to ethnical, cultural, and economic reasons causes inequality especially for female children. Especially the female children from lower social classes in countries without democratic development properties can experience social inequality more.

Chapter 2
The Influence of Changing Paradigms on Educational Management and School Administration

Şefika Şule Erçetin
Hacettepe University, Turkey

Ssali Muhammadi Bisaso
Islamic University in Uganda, Uganda & Hacettepe University, Turkey

ABSTRACT

In this study, an attempt was made to explore and examine the influence of changing paradigms on educational management. In approaching the chapter, the authors discussed the status-quo by describing the traditional management paradigm. The drivers to change which triggered off the rise of a new paradigm were highlighted and these dully led to a description of the new paradigm whose principles and theories were fully examined. The chapter then engaged a detailed analysis of the various changes in the education management and school administration field consequent to the shift in paradigms. The authors feel that an understanding of the major paradigms in education as well as those critical to educational management and administration is a universal calling. There is no gain saying, therefore, that this chapter will go a long way into improving educational management and administration both at structural level and at the school level.

DOI: 10.4018/978-1-5225-0460-3.ch002

INTRODUCTION

With the dawn of the 21st century, an unprecedented wave of change has become part of the reality we live with. The days of mass production as well as standardized products do not only appear to be over but are being replaced swiftly by quality production (Jamali, 2004). Accordingly, the key issues for the future are variety, flexibility, and customization of whatever we do. This rising wave of change has not spared the field of education as well. The sweeping volcano has meant that the old policies and systems can no longer obtain within new dispensations at all and thus call for redress at all levels and times. Managers of educational systems and school leaders should be on the lookout as a consequence.

Accordingly there has been a paradigm shift: this paradigm is reflected within experiences, beliefs, and values that affect perceptions. According to the Wikipedia, a paradigm involves the following aspects:

- Reasoning,
- A world view,
- Deeply held thinking,
- Anonymity and less awareness.

Because paradigms affect our view of what is real and true, we are slow to change them. Indeed, without paradigms, our minds would constantly suffer from sensory overload. Our minds use paradigms to filter all sensory input.

Viacana & Pedrozo (2010) contend that the "adaptive decision perspective adopted by post-normal science encourages the incorporation of human characteristics and potentialities that are currently neglected in management sciences". Accordingly, such characteristics and potentialities may imply evolving from an understanding of the human mind: the direction in this case should be from an automatic data processor to a logic that allows us to consider the interaction between emotion and cognition as well as between these two and their environment, thus connecting individual-organization-society. Nevertheless, the view points to the need for proper conceptualization of paradigm shifts that equally influence understanding of key scientific issues today.

For a proper analysis of the influence of changing paradigms on educational management and school administration, it is imperative to reduce the analysis to the following aspects:

- A description of paradigms,
- Relevance of paradigms,
- A description of paradigm shifts,

- Changing paradigms in educational management,
- The effect of the changing paradigms on the elements of educational management and school administration,
- Barriers to effective change within the management and administration realm.

PARADIGMS AND THEIR RELEVANCE

It is worth asking what a Paradigm is; and in attempting an answer to the same question, we argue that the word paradigm originates from the Greek paradeigma, para = "to place alongside"; deigma = "to show." It means to show by placing alongside, as in an Example, Pattern or Model. Many authors have attempted a definition of the term paradigm (Jamali, 2004; Rosaldo, 1996). Thus to Rosado (1996) a paradigm is "a mental construct, or conceptual model, influenced by our socialization, which defines and delimits the way we perceive reality and is the basis of our worldview". It is thus a particular way in which we things. Meanwhile Jamali (2004) states that a paradigm is "a framework of basic assumptions, theories and models that are commonly and strongly accepted and shared within a particular field of activity, at a particular point in time".

Essentially therefore, drawing from a multiplicity of descriptions, it is safe to say that a paradigm includes a set of assumptions which govern our interaction and interpretation of the world. Accordingly Jamali (2004) and Rosaldo (1996) it is important to recall that:

- Every human has a personal paradigm,
- A paradigm is influenced by outside forces,
- Personal experiences support a paradigm,
- Cultural situations influence paradigms,
- A person's status is critical to a paradigm,
- A paradigm is based on assumptions that are assumed to be true,
- Assumptions about paradigms cannot be fully tested.

Often the types of assumptions included in a paradigm are "tacit and arbitrary"; in other words, the people in the paradigm "don't consciously think about their existence or the potential for deeper meaning behind them" (Jamali (2004).

The purpose or relevance of a paradigm cannot be over emphasized since contemporary literature has dully focused on this. An over view of our global educational scenario and events in modern educational technology warrant the shift in paradigms (Quisumbing, 2001). The purpose of this paradigm shift can be viewed from the following perspective:

- Creating a basis to begin research,
- Narrow the amount of possible theories for observed phenomenon by "rejecting those that do not work in the paradigm,"
- Setting up the ground rules to evaluate new theories and ideas.

Huntington (1996) claims that simplified paradigms or maps are "indispensable for human thought and action". On the one hand, we may explicitly formulate theories or models and consciously use them to guide our behavior. Alternatively, we may deny the need for such guides and assume that we will act only in terms of specific "objective" facts, dealing with each case "on its merits." Based on Quisumbing (2001) and Huntington (1996), there is an "urgent call for explicit or implicit models" that can enable us:

- Generalize about reality,
- Understand causal relationships among phenomena,
- Anticipate and predict future developments,
- Distinguish what is important from what is unimportant,
- Show paths to be taken to achieve goals,
- Guide expectations,
- Sort, organize, and classify information,
- Affect types of questions asked,
- Incorporate accumulated knowledge, experience, and beliefs.

In the end, if the paradigm is successful in generating good ideas, it will even "generate the next paradigm that will replace it" (Quisumbing, 2001). However whenever the paradigm falls short of ability to explain present phenomena then its applicability is left wanting and thus becomes undesirable when it comes to defining key issues of the times. This paves way for the evolvement of another paradigm.

A DESCRIPTION OF PARADIGM SHIFTS

A new paradigm is normally generated whenever events occur but cannot be explained by the current paradigm. This would accordingly represent a paradigm shift. Rosado (1996) defines a paradigm shift as "a movement away from old explanations and structures that no longer explain reality, resulting in a redefinition of taken for granted boundaries due to the emergence of a new model or paradigm, which returns everyone back to zero". To him therefore a paradigm shift reflects:

- A different or unique mindset,
- A rise to a new way of viewing the world,
- Is inclusive of all views,
- Brings a change in societal and institutional structures,
- Creates an environment inclusive of all groups,
- Is safe for differences, and
- Everyone can benefit from it.

In the Structure of Scientific Revolutions, Kuhn wrote that "Successive transition from one paradigm to another via revolution is the usual developmental pattern of mature science." (Wikipedia). Paradigm shifts tend to appear in response to critical anomalies or an accumulation of them as well as the proposal of a new theory with the power to encompass both older relevant data and explain relevant anomalies.

A wide range of views and analyses (Huntington, 1996; Robinson, 2008; Quisumbing, 2001) indicate that paradigms can exist anywhere and indeed Kuhn applies this concept to the realm of scientific inquiry. He argues that "Western science has undergone numerous paradigm shifts, otherwise known as scientific revolutions. These events are triggered by a scientific theory so well proven and revolutionary that it changes the entire set of assumptions on which the current paradigm is based and is replaced by another set". This process, it should be noted, does not just happen instantly. It also takes a considerable amount of time for a scientific paradigm to change since they usually endure an extended period of influence. With modern technology however comes a "faster communication between scientists", this largely escalates the pace of appearance and collapse of paradigms.

According to the Wikipedia, "Social scientists have adopted the Kuhnian phrase 'paradigm shift' to denote a change in how a given society goes about organizing and understanding reality. A 'dominant paradigm' refers to the values, or system of thought, in a society that are most standard and widely held at a given time. Dominant paradigms are shaped both by the community's cultural background and by the context of the historical moment".

According to the Wikipedia, conditions that facilitate a system that can be accepted as a dominant paradigm are the following:

- Professional organizations that define its legitimacy,
- Leaders with dynamism to introduce and purport it,
- Journals and editors to write and disseminate essential information about the new system of thought,
- Government agencies for authoritative credence,
- Educators to propagate the paradigm's ideas through teaching students,

- Conferences devoted to discussing the paradigm,
- Media coverage,
- Lay groups with the belief in the paradigm,
- Sources of research funds for the paradigm.

It is ironic, therefore, that higher education is appearing more interconnected than ever, yet deep divisions exist over how best to conduct the core mission of education (Robinson, 2008). But what is crystal clear and worthy noting is a developing schism (divergence of views) within education between traditionalist groups and progressives. Huntington (1996) thus presents the following aspects of this argument.

- *Traditionalists* see the path to the future through the past. Here an ideal education system should reserve and inculcate values of the past.
- *Progressives* view learning-enabling technologies as part of education's evolution. Not only do they consider education a science, they believe in new issues playing a critical role in "transforming education" at all times and levels.
- *Outliers* elect to remain outside the fray. They steer clear of sides for they are not sure whether "innovations in education are for better or for worse".

However, despite the desire for reconciliation of the divergent orientations put forward above, the trends of formalized learning, and regulatory reform could indicate the early onset of paradigm change. Rather, such divisions are "precursors of fundamental change in educational practice" (Johnson, 2008; Keitiline et al, 2010). They equally signal several paramount shifts from:

- Intuition to science,
- Place to process,
- Time to competency,
- Handcrafted instruction to the formalized learning,
- Dispositional to structural learning,
- Undifferentiated instruction to personalized learning.

Given that "teaching and learning are complex, evolving practices", writes Keitiline et al (2010), it should be neither surprising nor unwelcome when disagreement emerges over their nature and purpose. Indeed, instead of being "divisive and adversarial", such conflict can serve as:

- An opportunity to reflect on our commitment to a particular ideological path.
- Identify a path to pursue which is useful.

- Clarifying the educational destination were headed for: whether desirable or simply a dead end.

This view is equally shared by Credaro (2006) and Jamal (2004). Since the paths of traditionalists and progressives seem to represent a fork in the road leading in divergent directions, the implication is that we must choose which of the two to follow, failure to do so thus, calls for an entirely new trail. That trail is the paradigm shift that will dictate undertakings in our times.

CHANGING PARADIGMS IN EDUCATIONAL MANAGEMENT

Education has undergone a paradigm shift in recent years. Systems that obtain in the education field are no longer the way they were and they should not be indeed given that we are living in changing times. "Education needs to undergo a radical paradigm shift in order to produce a super class of creative workers prepared to lead the new millennium economy, says an international proponent of the movement" (Huntington, 1996). When subjected to educational management, the story even gets more intriguing. In fact, Li Yan-zhong (2005) argued that an administration system is a system of complexity and chaos; the insensible disturbed motion in operation process sometimes can give rise to a dramatic fluctuation yet Credaro (2006) contends that if an administrative system lacks a proper corresponding mechanism, it must be subject to the harm of systematic deficiency.

Premised on the above backdrop of change, the field of management and administration has suffered some "degree of dislocation" (Jamali, 2004:104 citing Collins, 1996). This dislocation according to Jamali (2004) has in turn "catalyzed some soul-searching on the part of managers and academicians alike, and a reassessment of traditional managerial concepts and practices". It is this deficiency therefore, that leads to the rise of new paradigms with new techniques and initiatives to achieve transformation.

For nearly 200 years, the theories of Fayol, Taylor and Weber also known as "functional hierarchical line management" was the main management paradigm. The system was based on the view that:

- The management environment is largely stable,
- Centralized decision-making processes are more effective,
- Hierarchical communication channels must be instituted,
- Organizations are rational entities,
- Specific rational goals are ideal,

- Organizations should be highly formalized,
- Differentiated but efficient structures must exist.

According to Jaffee (2001) cited by Jamali (2004:105), this mechanistic orientation dominated most businesses in the past and is still "commonly encountered especially in the context of developing countries" but largely archaic in the developed world.

On their part, Garmston and Bruce, (1995) opine that "information from quantum mechanics, chaos theory, fractal geometry, and the new biology can help educators rethink school-improvement approaches". Interestingly, Erçetin & Bisaso (2015) have proposed a "fractal based leadership model" as a transition from the traditional management and leadership paradigm. The model considers critical challenges engulfing the "management terrain" today and offers a platform for "effective and efficient school management". Chaos as well as order exist simultaneously. Indeed issues like Crisis, Social, media-related and technological changes in the environment of education are now a visible phenomenon of this generation. Along the same argument, Erçetin, Bisaso & Saeed (2015) have opined that "understanding the chaos and complexity in educational systems" is made easy by conceptualization of "Fractal properties".

Issuing from the above premise it is imperative to analyze the major drivers of paradigm shifts or change. These according to Chapman (2001), Jamali (2004), Liyanage and Poon (2002) and Stallings (2000) are the following:

- A rapidly changing techno-socio-economic environment creating a need to diffuse information and technology within the organizations: this has proved to be beyond the capacity of the "old rigid hierarchal management system".
- Knowledge workers willingness to take initiative and responsibility.
- Employees are becoming more concerned about merit, value, worth, meaning and fulfillment.
- Consumption behavior reflected in expenditure is full of enlightenment, sophistication, inquisitiveness as well as critique.
- Globalization of society has produced an imperative for continual reappraisal of practices in order to maintain a competitive edge.

Looking at the new management paradigm, Jamali (2004) contends that organizations have become increasingly aware that the world has turned on its axis, necessitating a fundamental re-assessment of objectives, operations and management orientation. Therefore the 1980s have witnessed the emergence of a paradigm shift, or to be more accurate the search for new more appropriate paradigms.

Meanwhile, "contemporary management thinking" has been influenced by a number of theories. These create new insights into thinking about management processes. The theories in question are: behavioral approach, systems approach, contingency approach, culture excellence approach and organizational learning approach.

According to Jamali (2004) therefore;

- The behavioral approach focusses on the human factor and the importance of group dynamics and complex human motivations within the organization.
- The systems approach impresses upon managers the notions of interdependencies embeddedness within a system.
- The contingency approach underscores the value of adaptability or situational appropriateness.
- The culture-excellence approach reminds managers to accord more focus and attention to the softer issues of people, values, and employee or customer satisfaction.
- The organizational learning approach emphasizes usefulness of new knowledge and its applicability.

Reflecting on the base of the new management paradigm, Carnall (2003 cited by Jamali, 2004:108) contends that Management in the 21st century has accordingly taken a new orientation. He thus underlines the following key aspects worth noting:

- It is increasingly founded on the ability to cope with constant change and not stability,
- It is organized around networks and not hierarchies,
- It is built on shifting partnerships and alliances and not self-sufficiency,
- It is constructed on technological advantage and not bricks and mortar,
- It is based on virtual integration rather than vertical integration,
- It focuses on interdependence rather than independence, and
- Dwells on mass customization rather than mass production.

Related to the foregone view, Credaro (2006) argues that when considered from the perspective of education, this may be interpreted as "the need to update practices in keeping with the findings of international research, and to continually conform to national trends". This becomes even more challenging given the state of globalization and perspective to which all aspects of life, especially education, owe their abode.

It is however Sığrı (2010) who brings out the true picture of the paradigm shift in educational management when he presents three distinct categories as follows:

- The educational management paradigm of the past "symbolized by classical class environment" was only based on "the transfer of the knowledge" which describes the "interaction among instructors, learners and education/training contents viewed from educational/teaching traditions of the past".

- The educational management paradigm of today "symbolized by case studies" requires "a two-way communication between instructor and learners as well as among learners themselves".
- The educational management paradigm of the future "symbolized by distributed learning" requires a "knowledge base" (also coined as a learning resource center) which is placed on internet, and the "instructor's role as well as the learners' role is equally important for the learning process".

He thus concludes that the "educational management paradigms" of past and today primarily have focused upon the "interaction between instructors, learners and given subject matters". Naturally, the developments and structures in the surrounding society play an important role in determining the transition from one model to another.

THE INFLUENCE OF CHANGING PARADIGMS ON THE ELEMENTS OF EDUCATIONAL MANAGEMENT AND SCHOOL ADMINISTRATION

To dissect the effect of changing paradigms on educational management and administration especially as far as schools are concerned, it is imperative to explore this from the perspective of key elements of educational management and administration. We thus sought to do so by reducing the effect to the following pertinent elements:

Management structure: There is a shift from "cellular to matrix organizational structures" with fewer layers of management. This is in direct contrast to the old inflexible "multi-layered vertical hierarchical organizations".

Management philosophy: The shift is embracing "innovation as a key ingredient of success" as well as entrepreneurship and increased competitiveness in the management system as opposed to maintaining the status-quo. There is no fear of the unknown in this case.

Power structure: In the new dispensation there is encouragement of the proliferation of autonomy among the employees as opposed to autonomy of a selected few. Here power is shared and all members in an organization feel part of it.

Management approach: The shift reflects an "effort of harnessing people's creativity and enthusiasm" as opposed to conditioning them to "fixed policies". Whereas organizational goals are scared and have to be respected, when people in the organization are allowed to bring on board developmental ideas, the organization and management stand to benefit more.

Major management asset and treatment of labor: In the new management paradigm people are treated as the natural resource and capital asset of the organization and the most important source of "sustainable competitive advantage". This thus points to a high level of creativity within the organization. Here labor is also nurtured instead of being simply bought (see Table 1).

Meanwhile Castro (2012:338) observes the changes in the paradigms and their influence on management as follows:

Power limitations: In the new management paradigm there is focus on decentralization of power and authority as opposed to the centralized system. A key issue in new management therefore is the "distribution of power and authority"

Power forms: the new management paradigm calls for "redistribution of power between academic governance and managerial governance". This separation of administrative and academic matters has brought about a clear distinction between those involved in the two different areas. Meanwhile this ends up solving a plausible collision course.

In keeping with the same view and flow, Castro (2012:337) offers an extensive analysis on the nature of university management in the wake of a paradigm shift thus: there is basic responsibility to ensure that there is a balance between institutional autonomy and public accountability and there is also promotion of changes in the internal structure of universities

Implementation of university reforms alludes to a model change as discussed above. The model calls for "transfer of power from academics to bodies comprised of managers and external stakeholders". Such educational management organizations do a better job than the school level appointed officers due to the high level of accountability attached coupled with the expertise that comes with being trained to handle a specific task.

Table 1. Comparison of the traditional and new management paradigms

Traditional Management Paradigm	New Management Paradigm
Primary focus of management is reduction of costs of production	Reducing the indirect costs and improving competitiveness
Analysis of operations is based on a stability principle	Continuous improvement in flexibility and agility of operations
Managers regarded as decision-makers and labor as passive followers of instruction	Managers regarded as coaches/facilitators and labor as knowledge workers/intellectual capital

Source: Adapted from Khalil (2000) cited by Jamali (2004:109)

Relatedly Iancu et al, (2012:174) contends that three important changes have been realized with a shift in management paradigm:

- There is more focus on organizational networking and negotiation,
- Communication with the community has been given more importance, and
- Establishment of a structure competent for creating an effective public image has taken center stage in management.

On his part though, Demchenko (nd) opines that in the general trends and problems of Educational development in conditions of emerging new technologies and Global Information Environment, there are "new principles of Education in Information Environment based on an existing paradigm shift in modern education". He identifies three critical ones though:

- Research: this has become the biggest challenge to all educators in the present rapidly changed education environment caused by "Information Revolution."
- Teacher's role: there is a changing role of a teacher from "source to producer" of information. In fact when considered from the perspective of a knowledge society, in which education and teachers survive today, the role of a teacher has dully been transformed by the paradigm of knowledge creation and custodianship.
- Student's role: the shift sees "transformation of students from passive information receivers to active knowledge accumulators". Equally, the new paradigm sees students as having the potential to even learn on their own, which reduces the work of a teacher to a mere guide and facilitator of learning.

According to Demchenko, the new management paradigm therefore revolves around teamwork, participation, and learning. It also revolves around improved communication, integration, collaboration, and closer interaction and partnering with customers, suppliers and a wider range of stakeholders. Value creation, quality, responsiveness, agility, innovation, integration and teaming are increasingly regarded as useful guiding principles in the evolving new environment.

Credaro (2006) states that internal to the school are the pressures brought to bear by curricular reform. In keeping with Credaro (2006), therefore, educational institutions themselves must restructure the framework of their organisation to form "learning communities rather than institutions whose core function is the dispensing of information". The innovations and changes are at system, whole school and classroom level. Iancu et al., (2012) identify two key aspects: the paradigm of learning environments and learning activities and the paradigm of instructional resources which are not only institutional, but also affecting the time and energy of students.

Equally Neave (1998) cited by Castro (2012:337) argues that the general acceptance of this paradigm shift has been accompanied by the introduction of "new assessment procedures and models of accountability" in school systems.

Erçetin and Bisaso (2015) meanwhile opine that educational management and school administration obtains within a chaos and complex paradigm that calls for a shift from the traditional model of administration, the top-down hierarchy to a more flexible model based on fractals. This fractal based model, argue Erçetin and Bisaso (2015) would allow all members of the school system to partake in issues of school management and administration since they would be consulted.

Relatedly, understanding that education is born of chaos and complexity would also enable school managers to use fractal properties in conceptualizing school issues as well as taking decisions that inform practice (Erçetin, Bisaso & Saeed, 2015). To them therefore, what may appear as chaotic and complex in the school system, may actually be an opportunity to transform the same system. The new management paradigm based on understanding and conceptualizing fractals thus, makes us realize that within chaos and disorderliness, appears to be a high level of order and organization if well harnessed (see Table 2).

Managers also need to be abreast with the changing paradigm of teaching since management and administration obtain within supervision (Glanz, 2000; Gordon, 1992) done on teaching and learning processes. This "new paradigm of teaching" obtains within theory and condones research with clear application to instruction. In the "new paradigm of teaching", students are obliged to make an effort to reconstruct, discover and transform knowledge values through extension of academic works. The teacher's effort in this "new paradigm of teaching" is directed towards creating interaction between the student and learning materials as well as exchange of ideas between the student and the teacher (Johnson, Johnson, & Holubec, 1998). This completes both a bi-polar and tri-polar system (see Table 3).

FUTURE TRENDS

Paradigm shifts in educational management have and shall always have far-reaching effects and call for diverse measures in the education system. Learning management systems to develop new ways of dealing with student-related processes that can transform the entire education system have to be conceived.

Globalization should be embraced in order to establish how the melting of barriers among nations and their increasing interconnectedness, accelerated by technology, has led to a change in the world order its profound impact on global business.

Table 2. Summary of major changes in educational management and school administration

Element	Old Paradigm	New Paradigm
Understanding of Management	No need for change	Chaos, complexity and change
Organizational theories	Classic and mechanist	Modern
Organizational approaches	Organizational goals	Individual interest
Organizational structure	Pyramid	Flat and distributed hierarchies
Management structures	Top bottom	Interaction and flexibility
Management model	One size fits all	There is a mode for every size
Management behavior	Ascertained and controlled	Dynamic and situational
Understanding of Leadership	Certainty	Uncertain un-imaginable
Understanding of Education	Product focused	Process focused
Supervision structure	Frequency	Efficiency and effectiveness
Stability situation	Static	Non-static/dynamic
Conception of the future	Certain and definable	Uncertain and un-definable
Research approaches	Answer question of 'what' and more quantitative	Answer question of 'why' and more qualitative
Administrative tasks and duties	Generalizable	Relative and specific
Approach to information	Follows a clear pattern	Flows from all angles/directions
Systems related with management	School and authority based systems only	All social systems
Leadership characteristics	Work focused	Employee focused
Personnel management or leader and follower relations	Authority based	Family and friend
School type	Student Control center	Free Learning center
Content of school programs	Knowledge values	Values and attitudes
Perception of school individuals	Leaders and the led	Stakeholders and owners
School's view of the community	External to the school	Continuation of the school (annex)
Forms of school management	Policy based	Environment based

Source: Developed by the authors based on various readings

Technology which is the indisputably most critical facilitator behind the most far-reaching and powerful changes in business needs a proper system of harnessing as a way of transforming teaching, research and outreach programs.

Sustainability and Corporate Social Responsibility to improve on profitability at all levels and among all individuals in the organization ought to be conceived. The Study of Psychology to effect probing into cognition, motivation, behavior and

Table 3. Comparing the old and new teaching paradigms

Aspect	Old Teaching Paradigm	New Teaching Paradigm
Knowledge	Transferred from teachers to students	Joint venture based on sharing ideas
Students' role	Passive vessel and consumer	Active constructor of knowledge
Teachers' role	Sort and classify students	Aid development of students' competencies, skills and talents
School relationships	Impersonal	Personal interactions
Teaching context	Individualistic and competitive	Cooperative learning with teams on board
Teaching assumptions	Any expert can handle	Complexity of teaching calls for specialized training

Adopted from Johnson and Holubec, 1998 with modifications by the authors

performance - has become a key pillar of organizational management and more so, educational management.

Research on the possibility of the above trends within the education system and how they can be sustained should be the concern and focus of all stakeholders in the education system at all levels and times.

CONCLUSION

Effective change to any "organizational structure, philosophy or practice" is not mean feat and certainly not easily obtainable, a goal. This is premised on the view that at all levels one is likely to encounter challenges which turn out to be dynamic. The education area is becoming now the place where many technologies meet, converge and are tested by real practice and coexistence. Education becomes a conductor and promoter of new technologies and an area where new approaches and new social imperatives are forming now.

However whatever happens in education has a direct relationship with the management of education and school administration. Accordingly, the changing paradigms must be of major concern to the field of educational management as they ultimately influence the overall developments in the education system. Important to note though is that a paradigm shift is naturally inevitable and cannot be avoided in any case. Most importantly though, a shift in paradigm has far reaching consequences for educational management and thus all educational managers ought to be braced for the challenge at all times. This paper has explored the concept of paradigm, paradigm shift, relevance, key paradigm shifts, paradigm shifts in education and how such paradigms have influenced educational management as well as school administration. There is no gain saying, on the part of the authors, that the paper does not only add immensely to the growing body of knowledge in the social sciences but positions itself as an asset to stakeholders in the education system, most importantly managers and leaders.

REFERENCES

Credaro, A. (2006). Innovation and Change in Education. *Worrier Librarian*. Retrieved from www.warriorlibrarian.com/LIBRARY/innovate.html

Demchenko, Y. (1998). Paradigm Change in Education in Conditions of Emerging New Information Technologies and Global Information Infrastructure Building. *ED-Media/ED-Telecom 98 Proceedings*.

Erçetin, Ş. Ş., & Bisaso, S. M. (2015). The incorporation of fractals into educational management and its implications for school management models. In Ş. Ş. Erçetin & S. Banerjee (Eds.), *Chaos, Complexity and Leadership 2013*. Springer Proceedings in Complexity. doi:10.1007/978-3-319-09710-7_4

Erçetin, Ş. Ş., Bisaso, S. M., & Saeed, F. (2015). Understanding chaos and complexity in education systems through conceptualization of fractal properties. In Ş. Ş. Erçetin & S. Banerjee (Eds.), *Chaos, Complexity and Leadership 2013*. Springer Proceedings in Complexity. doi:10.1007/978-3-319-09710-7_12

Garmston, R., & Bruce, W. (1995). Adaptive schools in a quantum universe. *Educational Leadership*, *52*(7), 6–12.

Georgeta, I., & Castro, D. (2012). Governance in Spanish universities: Changing paradigms. *Journal of Hispanic Higher Education*, *11*(4), 336–350. doi:10.1177/1538192712441798

Glanz, J., & Linda, S. B. (2000). *Paradigm debates in curriculum and supervision: Modern and Postmodern. Key Education Books*. Westport, CT: Bergin and Garvey.

Gordon, P. (1992). Perspectives and imperatives: Paradigms, transitions and the new supervision. *Journal of Curriculum and Instruction*, *8*(1), 62–76.

Iancu, D. (2012). Educational management, a new paradigm in social development. *International conference of scientific paper afases*.Brasov.

Jamali, D. (2004). *Changing management paradigms: Implications for educational institutions. Suliman S.* Beirut, Lebanon: Olayan School of Business, American University of Beirut.

Johnson, D., Johnson, R., & Holubec, E. (1998). *Cooperation in the classroom.* Boston: Allyn and Bacon.

Quisumbing, L. R. (2001, January 10). *Changing paradigms for teacher education for the new millennium.* Paper read at the International Conference on Teacher Education, Philippine Normal University of Manila.

Robinson, K. (2008). *Changing education paradigms.* RSA.

Rosado, C. (1996, December 15). *Restructuring education for the 21st century.* Department of Urban Studies, Eastern University.

Sığrı, U. (2010). The paradigm shift in educational management: An evaluation of distributed learning as future approach. *The Turkish Online Journal of Educational Technology*, *9*(2), 137–149.

The Need for Paradigms and Models: Excerpted from Samuel Huntington. (1996). The New Era in World Politics. In *The Clash of Civilizations and the Remaking of World Order.* Retrieved from http://www.washingtonpost.com/wp-srv/style/longterm/books/chap1/clashofcivilizations.htm

Viacava, K. R., & Pedrozo, E. A. (2010). Higher education in management: Re-inventing the paradigm to gain the capacity to handle today's complexity. *On the Horizon*, *18*(1), 45–52. doi:10.1108/10748121011020994

Yan-zhong, L. (2005, December7). Fractal administrative organization: New exploration on administrative organizational pattern.*Workshop on Innovations in Governance and Public Service to achieve a Harmonious Society.*

KEY TERMS AND DEFINITIONS

Administration: The act of managing duties, responsibilities, or rules.

Chaos: Wikipedia defines the term chaos as any state of confusion or disorder while at dictionary.com it is defined as a state of utter confusion or disorder; a total lack of organization or order.

Complexity: The state of being complex or difficult with a variety of challenges and obstacles to effectiveness and efficiency. Complex situations are born of lack of a clear approach to dealing with an existing situation.

Leadership: The Oxford Advanced learner's dictionary of contemporary English views Leadership as a process of motivating, preventing problems, ascertaining policies, principles and objectives, training or teaching subordinates, showing the way, learning how work shall be done, believing in team spirit and adopting empathy through love.

Management: The act, art, or manner of managing, or handling, controlling, directing, etc.

Paradigm: This is a mental construct, or conceptual model, influenced by our socialization, which defines and delimits the way we perceive reality and is the basis of our worldview.

Paradigm Shift: The Wikipedia contends that a paradigm shift denotes a change in how a given society goes about organizing and understanding reality.

Chapter 3

Principal Proactivity: School Principals' Proactive Behavior Forms and Their Levels of Display

Tuncer Fidan
University of Ankara, Turkey

Ali Balcı
University of Ankara, Turkey

ABSTRACT

Increasing environmental complexity and ambiguity require organizations to rely on their members' proactive behaviors to deal with potential chaotic occurences surrounding the organizational functioning. Individuals respond these occurences by displaying various proactive behavior forms to build predictability islands in ocean of uncertainty. In this context, principal proactivity is one of key determinants of effectiveness of schools functioning in complex and anarchic environments. In this study, numerous forms of proactive behaviour were categorized under six general dimensions as innovativeness, issue selling, social networking, strategic scanning, strategic learning and feedback seeking. A causal comparative design was used to determine levels of general forms of proactive behavior performed by school principals. The data were collected from school principals working at different levels of public and private schools in İstanbul Province during 2014-2015 year. According to the results of study, the most frequent form of proactive behaviour displayed by school principals was strategic learning.

DOI: 10.4018/978-1-5225-0460-3.ch003

INTRODUCTION

The uncertainties led by the complexity and dynamism of organizational environments and the increasingly decentralized structures of organizations require the organizations to empower their members and extend the decision-making responsibilities throughout their organizational structures. A high level of environmental uncertainty brings limitations to the predictability of the direction of the change. Due to this uncertainty, organizations have increasingly been obliged to rely on the initiatives of their members to overcome the problems they face and to take advantage of the forthcoming opportunities. Instead of remaining reactive, individuals' performing proactive behaviors to deal with the opportunities and problems in internal and external environments of the organization has become an important determinant of organizational effectiveness (Crant, 2000; Russell & Russell, 1992; Frese & Fay, 2001).

Proactivity

According to Bateman and Crant (1993), the roots of proactive approach lie in the perspective of "symbolic interactionism". The theory of symbolic interactionism approaches humans as proactive and goal oriented beings (DeLamater, Myers, & Collett, 2015, 9). Humans do not only react to the actions of others, they also have the tendency to interpret or define the actions of others. The reactions they give are not targeted at others's actions directly, however, rather at the meanings they attribute to those actions (Blumer, 1986, 79). Meaning arises from interactions between individuals and individuals shape their perceptions towards events as a reaction to the interactions that they form with other individuals in a common context. According to this view, individuals contribute to the meaning arising from the situation in which they take part, and the individuals and the situation determine each other mutually (Reichers, 1987). In other words, individuals are the social products who are affected by the behaviors of others, but still keep the distance with others and have the ability to start an individual action (Blase, 1990). This interaction process enables individuals to make proactive behavioral choices based upon the shared meanings and to make action plans to implement their choices (Sass, 1994; DeLamater, Myers & Collett, 2015).

Thus, from the perspective of symbolic interactionism, employees can be considered as not passive recipients, but active individuals who attempt to shape their environments according to their expectations within organizational life. On this point, Crant (2000) defines proactive behavior as challenging the status quo to improve the current situation or to create a new one. In other words, proactive behavior is a means of changing something in a desired way (Bateman & Crant, 1999). According to Grant and Ashfort (2008) who adopt a more general perspective, a proactive behavior is an anticipatory action undertaken by employees to influence themselves or their environment.

Proactivity is not only to foresee the change, but also to create it. Proactivity necessitates taking initiative to ensure improvement in a task to be performed. Of course, not all behaviors leading to change can be considered as proactive. Behaviors which lead to unintentional change or those behaviors which only lead to change in perception but not in reality, can not be classified as proactive behaviors. In the light of such explanations, those who take an action, take new initiative and spark a constructive change and can foresee problems and opportunities can be defined as proactive individuals (Bateman & Crant, 1999).

The fact that the concept of proactivity has been discussed in the literature in association with the context-specific behaviors such as socialization, information seeking, etc. (Morrison, 1993; Wanberg & Kammeyer-Mueller, 2000), has led to various conceptualizations regarding proactive employee behaviors. Proactive employees are those who have become competant in their work, can show effective performance and can display problem-solving and organizational skills. They possess effective inter-personal qualifications, such as righteousness and leadership. They show a commitment to organizational goals and feel responsibility for sustaining the sucess and the values of their organizations. In addition, proactive employees have entrepreneurial characteristics such as taking initiative, having independent judgement and voicing their opinions fearlessly. Lastly, they display such characteristics as having personal integrity and organizing their own lives in accordance with higher values (Campbell, 2000; Salanova & Schaufeli, 2008). Crant (2000) states that proactive behaviors which may emerge in different forms consist of four common constructs:

1. **Proactive Personality:** This construct characterizes persons who can not be limited by the environmental powers and can have an impact on the environmental change, in comparison to other individuals. Proactive persons identify opportunities, take initiative, take actions and continue their actions until a significant change occurs (Bateman & Crant, 1993).
2. **Personal Initiative:** It can de defined as individuals' exceeding formal requirements by adopting the active and self-starting approach to the work goals and tasks. Personal initiative includes goal oriented and long-term behaviors embarked in accordance with the mission of the organization and sustained persistently by facing up to the obstacles standing in the way (Frese et al., 1997).

3. **Role Breadth Self Efficacy:** It is the confidence that individuals feel they can assume wider and more proactive roles than required by the technical tasks of their jobs. As this construct is concerned with a specific technical task (e.g. using a computer), it is different from the general self-efficacy concept (Parker, 1998).
4. **Taking Charge:** In order to ensure constructive changes, organizations need employees who are willing to challenge the status quo. In this context, taking charge can be defined as voluntary or constructive efforts of individuals to influence the organizational functions regarding the ways of performing tasks at the job, work unit or organizational level (Morrison & Phelps, 1999).

Although the four constructs suggested by Crant (2000) do cover the proactive behavior definition to a large extent on a conceptual level, a general consensus has not been reached on them yet. Discussions are still continuing on whether a proactive behavior explains anticipatory actions or not, or whether it is change oriented or not, or whether it is in-role or extra-role behavior, or whether the work environment variables have an impact on it or not (Grant & Ashford, 2008; Crant, 2000; LePine & Van Dyne, 1998). The use of proactivity to characterize both anticipatory and change-oriented behaviors, and both in-role or extra-role behaviors may be an appropriate solution for these issues (Grant & Ashford, 2008).

Forms of Proactive Behavior

Self-oriented and future-oriented behaviors such as implementing new work methods, influencing organizational strategy or learning new skills to deal with difficulties which may arise in the future can be characterized as proactive (Bindl & Parker, 2010). In light of this definition and discussions in the literature, it is possible to make mention of the existence of six general proactive behavior forms when organizational environment and work practices are considered (Balcı and others, 2012; Bateman & Crant, 1993; Dutton & Ashford 1993; Frese & Fay, 2001; Grant & Ashford, 2008; LePine & VanDyne, 1998; Morrison & Phelps, 1999; Parker & Collins, 2010; Wang et al., 2010):

1. **Innovativeness:** It can be described as employees' deliberately implementing or bringing new ideas, products, processes and procedures in their unit or organization. Innovative behavior comprises searching for new techniques, suggesting and implementing new ways and methods and providing the resources necessary to secure this process (Yuan & Woodman, 2010; West & Anderson, 1996). Innovative behavior starts with problem identification and

idea generation. Even if other employees do not agree, it is necessary to voice and suggest the new idea and find support for it. After these are ensured, the process is completed with the implementation of the new idea (LePine & Van Dyne, 1998; Scott & Bruce, 1994).

2. **Issue Selling:** These behaviors are performed by employees to draw the attention of other employees- notably high level managers- and influence their understanding of events, developments and trends which affect organizational performance (Dutton and others, 2001). Issue selling consists of discretionary and voluntary actions undertaken to shape the agenda of the organization. A set of strategic thinking and initiative actions is established by means of issue selling. Those which contribute to strategic behavior patterns are selected from this set and implemented (Ashford et al., 1998).
3. **Social Networking:** Proactive employees have a tendency to build a social environment which may help them to be successful in the workplace. These employees seek people who will support their personal initiatives and try to establish close relationships with powerful and influential people (Thompson, 2005). Employees in this network of relationships try to obtain the factors necessary for the personal and organizational success by means of information-seeking (Borgatti & Cross, 2003). Employees with high level of networking skills can easily make friends and create strong and beneficial coalitions. Moreover, these employees can place themselves in critical positions within the network to utilize existing opportunities and to create new ones (Ferris et al., 2005).
4. **Strategic Scanning:** Strategic scanning is an information acquisition activitiy carried out not for a problem encountered, but for exploratory purposes (Choudhury & Sampler, 1997). In strategic scanning every potential event, tendency or relationship in internal and external environments which may affect the functioning and the future performance of the organization is investigated and the information gathered is used in action plans which shape the future of the organization (Choo, 2001; Thomas, Clark & Gioia, 1993). Such investigations function as an early warning system and provide organizations with sufficient time and flexibility to deal with potential threats and opportunities. Proactive employees develop signals and indicators which inform of when problems or opportunities will arise. In turn, by the help of the information they gathered with such tools, they can identify the potential problems and opportunities before they occur, examine the advantages and disadvantages and develop alternative roadmaps, action plans and strategies (Frese & Fay, 2001; Preble, 1992).

5. **Strategic Learning:** Strategic learning is a learning behavior providing performance advantage and developing long-term adaptation capacity in a competitive environment. Strategic learning is set apart from other approaches to organizational learning due to the emphasis on learning for research, recognition and productivity (Thomas, Sussman & Henderson, 2001). No doubt, the most important component of strategic learning is information-seeking (Škerlavaj et al., 2007; Major & Kozlowski, 1997). Because, information seeking reduces the uncertainties and ensures the employee to understand, foresee and control the organizational environment. Further, by this means employees can obtain the information they require to ensure a higher level of performance and adaptation in the workplace (Morrison, 1993). From this perspective, it can be stated that strategic learning is a proactive information seeking and acquisition activity performed to prepare for potential future problems or opportunities (Choudhury & Sampler, 1997; Major & Kozlowski, 1997).
6. **Feedback Seeking:** Feedback seeking is the conscious efforts made to determine the accuracy and sufficiency of behaviors performed to reach the work goals (Ashford & Cummings, 1983). The concept of feedback is related to how the individuals' behaviors are perceived and evaluated by others. When feedback functions as a reward, it can increase the motivation of the employee. It may also be effective in improving the behaviors of employees in a desired way. Feedback seeking may occur in various forms. Employees may monitor the reactions of their co-workers and the clues they infer from the situations in which they take part. Or, employees may directly ask their co-workers how they perceive or evaluate them (Ashford, 1986).

Similar categorizations have been proposed in the literature. For instance, Grant and Parker (2009) state that proactive behavior forms can be classified into four groups: proactive work behavior, proactive person-environment fit behavior, proactive strategic behavior and proactive career behavior. However, Parker and Collins (2010) classified 11 different proactive behavior forms into three groups as proactive work behaviors, proactive person-environment fit behaviors and proactive strategic behaviors in their model in which proactive behavior constructs were presented as separate behavior forms. On the other hand, Crant (2000) argues that the interactions between proactive behavior constructs and environmental factors lead to work behaviors such as identifying opportunities, challenging the status quo, creating a better work environment, socialization, feedback seeking, issue selling and innovativeness. Similarly the objective of this study is to ascertain the levels of general forms of proactive behaviors performed by school principals. The measureable aims of the study are;

1. To determine to what extent school principals perform innovativeness, issue selling, social networking, strategic scanning, strategic learning and feedback seeking dimensions of proactivity.
2. To determine whether innovativeness, issue selling, social networking, strategic scanning, strategic learning and feedback seeking levels of school principals differ according to gender, professional tenure, managerial tenure and school type.

METHODOLOGY

Research Design

A causal comparative design was used as a descriptive survey in the study. In this scope, the proactive manager scale was administered to determine the school principals' levels of performing innovativeness, issue selling, social networking, strategic scanning, strategic learning, feedback seeking dimensions of proactivity. In addition, a comparison was made between proactivity levels of school principals according to various demographic characteristics.

Population and Sample

The population of the study consisted of 5249 school principals working in the 39 districts of the province of Istanbul. The sample size was determined as 358 participants from the theoretical tables (Balcı, 2015). In the study, "proportional stratified sampling" technique was adopted to choose the school principals working in the districts of Kadıköy, Fatih, Üsküdar, Küçükçekmece, Kartal, Ümraniye, Bağcılar, Bahçelievler, Sultangazi and Sultanbeyli as the sub-strata with respect to the "Istanbul Life Quality Index" developed by Şeker (2011) who classified 39 districts into five groups in terms of their levels of living standards. 400 copies of questionnaires were sent to the participants of the study (school principals) and 368 of them were returned. Of the questionnaires returned, 41 were excluded due to missing items, response errors and outliers. 327 questionnaires, amounting to 92% of the sample size, were included in analysis.

Data Collection Tool

The "proactive manager scale" used in the study was developed by the researchers by drawing from the literature on proactivity (Balcı et al., 2012; Bateman & Crant, 1993; Dutton & Ashford 1993; Frese & Fay, 2001; Grant & Ashford, 2008; LePine

& Van Dyne, 1998; Morrison & Phelps, 1999; Parker & Collins, 2010; Wang et al., 2010). In addition, two items were taken from a scale developed by Bateman and Crant (1993) and a further two items from the scale of Morrison and Phelps (1999). Necessary permissions were obtained as for the use of these items in the scale. The draft scale consisting of 39 items was sent to a group of experts including questionnaire development experts, researchers and practitioners from the related field and necessary revisions were made according to the feedback received. In order to test the construct validity of the scale, a factor analysis was conducted on the results obtained from the pilot study. 11 items with factor loadings less than .32 and one overlapping item were removed from the scale. Another further item with item total correlation less than .20 was also removed. As a result, it was found that the 26 items were loaded on six factors and these factors explain 60.85% of the total variance. The overall Cronbach alpha coefficient for all items is .88. The response format was standardized using five-point Likert scale ranging from (1) "Never" to (5) "Always". The items on the scale and their factor loadings, item total correlations and Cronbach alpha coefficients for each subscale were presented in the Appendix.

Data Analysis

In the analysis of data gathered via proactive manager scale, descriptive statistics such as arithmetic mean and standart deviation were used to determine innovativeness, issue selling, social networking, strategic scanning, strategic learning and feedback seeking levels of school principals. Median and Spearman's correlation were also used because the normal distribution was not obtained for issue selling dimension. To determine whether the levels of school principals in each proactivity dimension differ significantly in terms of gender, professional tenure, managerial tenure and school type independent samples t-test and one way analysis of variance were conducted, and Mann Whitney-U test and Kruskal-Wallis test were used in case the normal distribution was not obtained. Significance testing was performed at a two tailed α-level of 0.05.

FINDINGS

Table 1 shows a summary of descriptive statistics for the scores of the school principals (from here on refered as "participants") on the dimensions of proactive behavior.

According to Table 1, the most frequent proactive behavior form displayed by the participants is strategic learning ($\bar{X}$= 4.50). It is followed respectively by issue selling ($\bar{X}$= 4.37), innovativeness (4.29), feedback seeking ($\bar{X}$= 4.25), strategic

Table 1. Descriptive statistics regarding the proactive behavioral dimensions of participants

Dimensions	N	Mean	Standard Deviation	Median	1	2	3	4	5	6
1. Innovativeness	327	4.29	0.41	4.33	-					
2. Issue Selling	327	4.37	0.49	4.50	0.51*	-				
3. Social Networking	327	3.67	0.78	3.72	0.28*	0.25*	-			
4. Strategic Scanning	327	4.16	0.53	4.20	0.51*	0.42*	0.38*	-		
5. Strategic Learning	327	4.50	0.39	4.50	0.36*	0.22*	0.13*	0.35*	-	
6. Feedback Seeking	327	4.25	0.56	4.30	0.37*	0.21*	0.35*	0.29*	0.36*	-

* Spearman rho $p < 0.05$

scanning ($\bar{X}$= 4.16) and social networking ($\bar{X}$= 3.67). It can be seen in Table 1 that there is a low and medium level positive relationship between the sub-dimensions of proactivity. The highest level of correlation was found between innovativeness and issue selling ($r_s = .51$) and strategic scanning ($r_s = .51$) dimensions.

Statistical analyses regarding the proactivity levels of participants according to gender, professional tenure, managerial tenure and school type which were used as independent variables are presented below.

Investigation of Proactivity Levels of School Principals According to Gender

Independent samples t-test was used to compare the mean scores of participants on innovativeness, social networking, strategic scanning, strategic learning and feedback seeking subscales according to gender. Results are shown in Table 2.

Table 2 indicates that there is no significant difference between the mean scores of female and male participants on innovativeness, social networking, strategic scanning, strategic learning and feedback seeking subscales. The Mann-Whitney-U test results for issue selling subscale according to gender can be seen in Table 3.

According to Table 3, the resulting z value is not significant. No significant difference between the mean ranks of female and male participants on issue selling subscale was found. The results given in Table 2 and 3 show that there is no difference between the frequencies of proactive behaviors of female and male participants.

Table 2. T-tests results for the comparison of the mean scores according to gender

Dimensions	Gender	N	$\bar{X}$	S	df	t	p
Innovativeness	Female	84	4.25	0.49	118.36	-0.84	0.41
	Male	243	4.30	0.37			
Social Networking	Female	84	3.75	0.82	325	-0.95	0.30
	Male	243	3.65	0.75			
Strategic Scanning	Female	84	4.16	0.63	118.21	-0.19	0.85
	Male	243	4.17	0.48			
Strategic Learning	Female	84	4.55	0.37	325	1.33	0.18
	Male	243	4.49	0.39			
Feedback Seeking	Female	84	4.27	0.61	325	0.61	0.54
	Male	243	4.23	0.54			

Table 3. Mann-Whitney-U test results for the comparison of the issue selling scores according to gender

Dimension	Sex	N	Mean Rank	Rank Sum	U	Z	p
Issue Selling	Female	84	171.89	14439.00	9543.00	-0.90	0.37
	Male	243	161.27	39189.00			

Investigation of Proactivity Levels of School Principals According to Professional Tenure

Results of one-way analysis of variance conducted to test for the differences in the mean scores of participants on innovativeness, social networking, strategic scanning, strategic learning and feedback seeking subscales according to professional tenure can be seen in Table 4.

When Table 4 is examined it is revealed that there is a significant difference between the mean scores of professional tenure groups on feedback seeking subscale [$F_{(2,324)}$= 3.42, p< 0.05]. The participants with 26 and more years of professional tenure display feedback seeking behaviour more frequently than the participants with 15 and less years of professional tenure. The results of the Kruskal-Wallis test carried out for the issue selling sub-dimension is presented in Table 5.

Results of the analysis indicate that there is a significant difference between the mean ranks of participants on issue selling subscale according to professional tenure. [$X^2_{(2)}$ = 8.94, p<.05]. Post hoc testing with the Mann-Whitney U test demonstrates

Table 4. ANOVA results for the comparison of the mean scores according to professional tenure

Dimensions	Professianal Tenure Groups	N	$\bar{X}$	S	df	F	p	Significant Difference
Innovativeness	15 years and below	99	4.25	0.44	2.324	1.35	0.26	-
	16-25 years	113	4.27	0.40				
	26 years and above	115	4.34	0.37				
Social Networking	15 years and below	99	3.67	0.83	2.324	2.05	0.13	-
	16-25 years	113	3.57	0.73				
	26 years and above	115	3.78	0.75				
Strategic Scanning	15 years and below	99	4.11	0.61	2.324	2.58	0.08	-
	16-25 years	113	4.12	0.44				
	26 years and above	115	4.25	0.51				
Strategic Learning	15 years and below	99	4.49	0.43	2.324	0.02	0.98	-
	16-25 years	113	4.50	0.38				
	26 years and above	115	4.50	0.35				
Feedback Seeking	15 years and below	99	4.12	0.60	2;324	3.42	0.03	(Scheffe) 3-1
	16-25 years	113	4.26	0.57				
	26 years and above	115	4.32	0.51				

Table 5. Kruskal-Wallis results for the comparison of issue selling scores according to professional tenure

Dimensions	Professional Tenure Groups	N	Mean Ranks	sd	X^2	p	Significant Difference (U)
Issue Selling	15 years or below	99	141.61	2	8.94	0.01	2-1
	16-25 years	113	168.32				3-1
	26 years and above	115	179.03				

that the participants with 16-25 years and 26 and more years of professional tenure display issue selling behaviour more frequently than those with 15 and less years of professional tenure. It is seen that the frequency of the issue selling behaviour increases with the professional tenure.

Investigation of the Proactivity Levels of School Principals According to Managerial Tenure

Table 6 presents the t-test results comparing the mean scores of participants on innovativeness, social networking, strategic scanning, strategic learning and feedback seeking subscales according to managerial tenure.

From Table 6, it can be seen that there is no significant difference between innovativeness, social networking, strategic scanning, strategic learning and feedback seeking scores of the participants in regard to managerial tenure. Table 7 shows the Mann Whitney-U test results for the issue selling sub-dimension.

As seen in Table 7, the z value obtained from the Mann Whitney-U test is not significant. The results in Table 6 and 7 reveal that there is no significant difference between the frequencies of proactive behaviors of participants according to managerial tenure.

Table 6. T-tests results for the comparison of the mean scores according to managerial tenure

Dimensions	Managerial Seniority	N	$\bar{X}$	S	df	t	p
Innovativeness	9 years and below	171	4.26	0.42	325	-1.26	0.21
	10 years and above	156	4.32	0.38			
Social Networking	9 years and below	171	3.65	0.76	325	-0.62	0.54
	10 years and above	156	3.71	0.78			
Strategic Scanning	9 years and below	171	4.15	0.54	325	-0.54	0.92
	10 years and above	156	4.18	0.50			
Strategic Learning	9 years and below	171	4.48	0.41	325	0.78	0.44
	10 years and above	156	4.51	0.36			
Feedback Seeking	9 years and below	171	4.21	0.58	325	0.92	0.36
	10 years and above	156	4.27	0.54			

Table 7. Mann-Whitney-U test results for the comparison of the issue selling scores according to managerial tenure

Dimension	Managerial Tenure	N	Mean Rank	Rank Sum	U	Z	p
Issue Selling	9 years and below	171	155.92	26662.00	11956.00	-1.64	0.10
	10 years and above	156	172.86	26966.00			

Investigation of the Proactivity Levels of School Principals According to School Type

Independent samples t-test was conducted to compare the mean scores of participants on innovativeness, social networking, strategic scanning, strategic learning and feedback seeking subscales according to school type. Results are shown in Table 8.

Table 8 indicates that there is no significant difference between the mean scores of the participants on innovativeness, social networking, strategic scanning, strategic learning and feedback seeking subscales according to school type. The Mann Whitney-U test results for the sub-dimension of issue selling are given in Table 9.

As seen in Table 9, the resulting z value is found to be significant [Z= -2,30, p <.05]. Participants working at private schools display a higher frequency of issue selling behavior in comparison to their counterparts working at public schools. According to the results shown in Table 8 and 9, there is only a significant difference in the issue selling sub-dimension in terms of school type.

Table 8. T-tests results for the comparison of the mean scores according to gender

Dimension	School Type	N	$\bar{X}$	S	df	t	p
Innovativeness	Public	272	4.29	0.39	325	0.33	0.74
	Private	55	4.27	0.45			
Social Networking	Public	272	3.64	0.80	96.16	-1.95	0.55
	Private	55	3.83	0.60			
Strategic Scanning	Public	272	4.15	0.53	325	-1.19	0.27
	Private	55	4.24	0.48			
Strategic Learning	Public	272	4.50	0.39	325	-0.06	0.95
	Private	55	4.50	0.39			
Feedback Seeking	Public	272	4.24	0.56	325	-0.39	0.69
	Private	55	4.27	0.55			

Table 9. Mann-Whitney-U test results for the comparison of the issue selling scores according to school type

Dimension	School Type	N	Mean Rank	Rank Sum	U	Z	p
Issue Selling	Public	272	158.66	43155.50	6027.50	-2.30	0.02
	Private	55	190.41	10472.50			

DISCUSSION

The facts that the decision-making processes and financial accountability mechanisms have become more decentralized and competition between schools has been encouraged, and that discussions concerning school performance have become heated (Mulford, 2003), have led proactive school principals who are student-centered, thoughtful and sensitive to ethical values and have transformational leadership skills (Gupton, 1998, 21) to become more prominent. Discovering underlying individual and organizational causes of the proactive principal behaviors have been an important challenge for researchers approaching proactivity as a precursor of change, innovation and entrepreneural activities at schools (Eyal & Kark, 2004; Heilbrunn, 2010). Because it is stated that schools are seen as professional learning communities enabling principals to exhibit more creativity and entrepreneurship skills (Thompson and others, 2004; Giles & Hargreaves, 2006). For this reason, instead of a correlational approach in which the concepts or constructs affecting proactive behavior forms are included in a research, an exploratory approach was adopted to define general proactive behavior forms and to reveal the frequencies of these behavior forms. Moreover, a set of individual and organizational factors potentially having an impact on those behavior forms were examined in the final stage.

This study found out that there was no significant difference between male and female school principals in the frequency of displaying proactive behaviours. However, a review of the literature in this area yields conflicting findings regarding the impact of gender on innovativeness, issue selling and strategic scanning behaviors. For example, DiTomaso and Farris (1992) found that female R&D engineers perceived themselves to be less innovative compared to males, while Sonfield, Lussier, Corman and McKinney (2001) state that there was no difference between male and female business owners with respect to choosing innovation strategies. Similarly, Carli (2001) states that males can exert more social influence – and thus, they can sell issues more often- than females because of the group formation processes, prejudices and differences in communication styles whereas O'Neil (2004) found that gender was not a significant predictor of upward influence. Conflicting findings of the studies on issue selling deserve an attention as well. Othman and Hamedon (2011) indicate that the frequency of environmental scanning differs significantly according to gender. They further elaborate by stating that female managers undertook scanning activities more frequently than male counterparts. On the contrary, Miller (1994) purports that males display a higher frequency of environmental scanning behaviour.

The findings of this study concerning the differences between social networking, strategic learning and feedback seeking behaviours of female and male school principals, to a large extent, offer a replication of those reported in the literature. In general, no significant relationship was found between gender and social networking (Brass, 1985; Ferris et al., 2005; Blass et al., 2007), strategic learning (Downing and others, 2008) and feedback seeking (Renn & Fedor, 2001; London et al., 1999; Teunissen et al., 2009) behaviors in the studies in the literature.

Employees tend to display more proactive behaviors such as problem solving, idea implementation, prosocial rule breaking and role enlargement under the conditions of professional and organizational autonomy developed with experience (Grant & Ashford, 2008). Similarly, in this study the school principals with 16-25 and 26 and more years of professional tenure were found to display issue selling behavior more frequently than those with 15 and less years of professional tenure. At this point, Howard-Grenville (2007) points out that issue selling skills are acquired over time through formal authority, normative knowledge and experiences gained from failures and resistances encountered. In other words, the results of this study support the findings of Bishop, Webber & O'Neill (2011) who suggested that employees in managerial positions get more prepared for issue selling as their organizational tenures increase. The finding that there was no significant difference in issue selling behavior according to managerial tenure can be explained by the appointment of teachers to managerial positions in the later stages of their careers.

On the other hand, findings of this study did not match the general tendencies found in the literature in regard to the relationship between tenure and feedback seeking. According to the research findings, school principals with 26 and more years of professional tenure showed a higher frequency of feedback seeking behavior in comparison to those school principals with less than 15 years of professional tenure. This finding does not support the previous findings indicating that feedback seeking behavior of an employee decreases, as his or her tenure which implies increased role clarity and decreased uncertainty increases over time (Ashford, 1986; Miller & Jablin, 1991; VandeWalle et al., 2000; Barner-Rasmussen, 2003). The appointment of teachers to serve as school principals in the later stages of their careers may again account for the finding that no significant difference was found in feedback seeking behavior according to managerial tenure.

Conflicting research findings concerning the innovativeness and strategic scanning behaviors according to professional and managerial tenure have been reported frequently in the literature. For example, while Pelled (1996) stated that there is a positive relationship between professional tenure and creativity in work groups, Kimberley and Evanisko (1981) concluded that managerial tenure does not have a significant effect on innovativeness, which is similar to the findings of this study. Similar to the findings regarding innovativeness behavior, it was found in this study that strategic scanning behaviours of the school principals did not differ according to professional and managerial tenure. However conflicting findings are reported in the literature. There are findings showing that as managerial tenures of managers increase, their perspectives get narrower and their scanning activities become more limited (Keck & Tushman, 1993; Miller and others, 1996). On the other hand, Abebe, Angriawan and Tran (2010) and Hambrick, Cho and Chen (1996) found that organizational and managerial tenure has a positive significant effect on the frequency of environmental scanning.

The finding that professional and managerial tenures do not have a significant effect on social networking and strategic learning behaviors generally supports the results in the literature. For instance, Luo (2003), Juenke (2005) and McCallum et al. (2014) stated that there is no significant relationship between professional and managerial tenure and social networking. Similarly, Chiang and Wang (2008) and Wang et al. (2010) did not find a significant relationship between tenure and learning behavior.

It is indicated that autonomy regarded as one of the conditions of the emergence of a proactive behaviour (Grant & Ashford, 2008) is reduced and limited by employment regulations, rules governing relationships with employees, detailed and restrictive legal regulations, intervention of numerous stakeholders and strict organizational structures in public sector organizations such as schools (Perry & Rainey, 1988). Findings concerning issue selling are supportive of this argument. It was found in this study that private school principals sell issues more frequently when compared to public school principals. Similar findings have been reported in the literature. According to these findings, practices for influencing superiors in the private sector are seen more frequently in comparison to the public sector (Rainey, 1983) and private sector employees can make use of more strategies to influence their superiors (Hooijberg & Choi, 2001).

On the contrary, the findings concerning innovativeness, social networking, strategic scanning, strategic learning and feedback seeking behavior do not support the pro-private sector arguments in the literature. For example, Friedman (1997) states that competition leads to more rapid innovations in the curriculum and teaching and private schools can adapt to this process much more quickly than public schools can do. For this reason, it is claimed that private schools can be a better model for innovation in comparison to public schools. Moreover, private schools are more prone to the pressures of external standards requiring the principals to be more innovative (Coleman et al., 1982; Maranto, 2003).

Additionally, despite the bureaucratic tendencies and restrictive formal relationships within public organizations (Eglene et al., 2007), the social networking skills of managers and employees in both the public and private sector were generally found to be close to each other at above average level (Douglas & Ammeter, 2004; Li & Zhang, 2007; Todd et al., 2009; Brosky, 2011). Unlike the findings of this study, the discussions pertaining to strategic scanning in the literature have rather evolved to the detriment of public sector organizations due to the influences of numerous external stakeholders on public organizations, restrictive legislative regulations and control mechanisms (Hendrick, 2003). On the other hand, environmental strategic scanning is easier for private sector organizations, as they are sufficiently autonomous to determine their own scanning strategies (Gunn, 1988).

Strategic learning and feedback seeking are often regarded in the literature as concepts developed for private sector organizations (Bate & Robert, 2002; Rashman et al., 2009). In contrast to the findings of this study, Skule (2004) found out that the level of informal learning of employees in the public sector is lower than that of employees in the private sector. The reason behind this can be explained by the fact that the boundaries between the units, functions and processes within public organizations are structural barriers to learning and information flow (Newman et al., 2000). Hassan and Rohrbaught (2009) state that, due to such barriers; it is not always possible to obtain continuous, direct and useful feedback in public organizations. However, Abu-Doleh and Weir (2007) suggested that the difference between public and private sector organizations results from the content of the feedback. They remarked that the feedback in public sector is mostly related with the efforts to comply with legislative regulations, but not with individual performance.

In addition to the discussions presented, this study has been conducted on the population of school principals and it relies on the self reports of these participants. Different findings may be obtained in studies in which cross reports and different populations are used. Furthermore, data obtained from observers may furnish researchers with more refined and diversified perspectives. However, the fact that supervisors and co-workers may view the proactive behavior negatively since it is associated with challenging the status quo must be taken into consideration in studies conducted with third-party observers (Frese & Fay, 2001; Parker & Collins, 2010). Lastly, as the findings of this study are limited to the schools which are under the influence of both a social environment with a high level of complexity and uncertainty and a public sector environment driven by strict legal regulations (Giles & Hargreaves, 2006), additional studies must be conducted in order to generalize the results to non-educational organizations.

SOLUTIONS AND RECOMMENDATIONS

Despite the limitations of the study mentioned above, numerous forms of proactive behaviour were categorized in this study under six general dimensions as innovativeness, issue selling, social networking, strategic scanning, strategic learning and feedback seeking which reflect the properties of the constructs empirically identified by Crant (2000). Furthermore, a considerable number of studies regarding the forms of proactive behaviour have been conducted for non-educational organizations. However there are currently not enough empirical studies on the proactive behaviour forms of the managers of educational organizations. Thus, the issue of which proactive behaviours have been displayed at what frequency has potential to become an important subject of a theoretical discussion for future studies.

Similary, the individual and organizational level factors such as gender, professional tenure, managerial tenure and public-private school distinction were included in this study as it was realized that the studies in the literature concerning whether demographic and organizational factors have effects on the proactive behaviour forms of school principals are rare. In the future, research should focus on further studying the effects of these factors on managerial proactivity in different contexts and on additional control variables such as organizational culture and social structure that may have produced omitted variable problem (LePine & Van Dyne, 1998). Another task for future research is assessment of the inter-relationships among proactive behavior constructs and six general forms discussed in this study. The relationships between these behavioral forms and positive organizational outcomes (individual and organizational performance improvement, the increase in student success, etc.) should also been explored in detail in future research.

In conclusion, it was ascertained in this study that the most frequent form of proactive behaviour displayed by school principals was strategic learning, while the least frequent proactive behaviour displayed was social networking. In contrast to the abundant literature on feedback seeking, it was found that feedback seeking behaviour was performed more frequently as the level of professional tenure increased. However, the findings concerning issue selling were generally found to be congruent with those found in the literature. School principals with a higher level of professional tenure and those working in private schools displayed a higher frequency of issue selling behaviour. In addition to this, findings on the proactive behaviour forms displayed by school principals showed no significant differences in terms of gender and managerial tenure. Accordingly, these findings on one hand have revealed the proactivity levels of school principals in the school context, and on the other hand provide opportunities for future studies in which inter-relationships among constructs, forms and outcomes of proactive behaviour should be investigated in a model.

REFERENCES

Abebe, M., Angriawan, A., & Tran, H. (2010). Chief executive external network ties and environmental scanning activities: An empirical examination. *Strategic Management Review*, *4*(1), 30–43.

Abu-Doleh, J., & Weir, D. (2007). Dimensions of performance appraisal systems in Jordanian private and public organizations. *International Journal of Human Resource Management*, *18*(1), 75–84. doi:10.1080/09585190601068334

Ashford, S. J. (1986). Feedback-seeking in individual adaptation: A resource perspective. *Academy of Management Journal*, *29*(3), 465–487. doi:10.2307/256219

Ashford, S. J., & Cummings, L. L. (1983). Feedback as an individual resource: Personal strategies of creating information. *Organizational Behavior and Human Performance*, *32*(3), 370–398. doi:10.1016/0030-5073(83)90156-3

Ashford, S. J., Rothbard, N. P., Piderit, S. K., & Dutton, J. E. (1998). Out on a limb: The role of context and impression management in selling gender-equity issues. *Administrative Science Quarterly*, *43*(1), 23–57. doi:10.2307/2393590

Balcı, A. (2015). *Sosyal bilimlerde arastirma: Yöntem, teknik ve ilkeler. (10'th Edition)*. Ankara: PegemA Yayincilik.

Balcı, A., Baltacı, A., Fidan, T., Cereci, C., & Acar, U. (2012). Örgütsel sosyalleşmenin, örgütsel özdeşleşme ve örgütsel vatandaşlıkla ilişkisi: Ilköğretim okulu yöneticileri üzerinde bir araştırma. *Eğitim Bilimleri Araştırmaları Dergisi-Journal of Educational Sciences Research*, *2*(2), 47–74.

Barner-Rasmussen, W. (2003). Determinants of the feedback-seeking behaviour of subsidiary top managers in multinational corporations. *International Business Review*, *12*(1), 41–60. doi:10.1016/S0969-5931(02)00087-2

Bate, S. P., & Robert, G. (2002). Knowledge management and communities of practice in the private sector: Lessons for modernizing the National Health Service in England and Wales. *Public Administration*, *80*(4), 643–663. doi:10.1111/1467-9299.00322

Bateman, T. S., & Crant, J. M. (1993). The Proactive Component of Organizational Behavior: A Measure and Correlates. *Journal of Organizational Behavior*, *14*(2), 103–118. doi:10.1002/job.4030140202

Bateman, T. S., & Crant, J. M. (1999). Proactive behavior: Meaning, impact, recommendations. *Business Horizons*, *42*(3), 63–70. doi:10.1016/S0007-6813(99)80023-8

Bindl, U., & Parker, S. K. (2010). *Proactive work behavior: Forward-thinking and change-oriented action in organizations* (Vol. 2). Washington, DC: American Psychological Association.

Bishop, K., Webber, S. S., & O'Neill, R. (2011). Preparation and prior experience in issue-selling success. *Journal of Managerial Issues*, 323–340.

Blase, J. J. (1990). Some Negative Effects of Principals' Control-Oriented and Protective Political Behaviour. *American Educational Research Journal*, *27*(4), 727–753. doi:10.3102/00028312027004727

Blass, F. R., Brouer, R. L., Perrewé, P. L., & Ferris, G. R. (2007). Politics Understanding and Networking Ability as a Function of Mentoring The Roles of Gender and Race. *Journal of Leadership & Organizational Studies*, *14*(2), 93–105. doi:10.1177/1071791907308054

Blumer, H. (1986). *Symbolic Interactionism: Perspective and Method*. Berkeley, CA: University of California Press.

Borgatti, S. P., & Cross, R. (2003). A relational view of information seeking and learning in social networks. *Management Science*, *49*(4), 432–445. doi:10.1287/mnsc.49.4.432.14428

Brass, D. J. (1985). Men's and women's networks: A study of interaction patterns and influence in an organization. *Academy of Management Journal*, *28*(2), 327–343. doi:10.2307/256204

Brosky, D. (2011). Micropolitics in the School: Teacher Leaders' Use of Political Skill and Influence Tactics. *International Journal of Educational Leadership Preparation*, *6*(1), n1.

Campbell, D. J. (2000). The proactive employee: Managing workplace initiative. *The Academy of Management Executive*, *14*(3), 52–66.

Carli, L. L. (2001). Gender and social influence. *The Journal of Social Issues*, *57*(4), 725–741. doi:10.1111/0022-4537.00238

Chiang, T. L., & Wang, J. (2008). *Workplace Learning, Job Satisfaction, and Organizational Commitment in Small to Midsize Companies in Taiwan*. Online Submission, retrieved on 15.05.2015.

Choo, C. W. (2001). Environmental scanning as information seeking and organizational learning. *Information Research*, *7*(1), 7–1.

Choudhury, V., & Sampler, J. L. (1997). Information specificity and environmental scanning: An economic perspective. *Management Information Systems Quarterly*, *21*(1), 25–53. doi:10.2307/249741

Coleman, J., Hoffer, T., & Kilgore, S. (1982). Cognitive outcomes in public and private schools. *Sociology of Education*, *55*(2/3), 65–76. doi:10.2307/2112288

Crant, J. M. (1996). The Proactive Personality Scale as a Predictor of Entrepreneurial Intentions. *Journal of Small Business Management*, *34*(3), 42–49.

Crant, J. M. (2000). Proactive Behavior in Organizations. *Journal of Management*, *26*(3), 435–462. doi:10.1177/014920630002600304

DeLamater, J. D., Myers, D. J., & Collett, J. L. (2015). Social Psychology (8th ed.). Boulder, CO: Westview Press.

DiTomaso, N., & Farris, G. F. (1992). Diversity in the high-tech workplace: diversity and performance in R&D. *Spectrum, IEEE, 29*(6), 21-24.

Douglas, C., & Ammeter, A. P. (2004). An examination of leader political skill and its effect on ratings of leader effectiveness. *The Leadership Quarterly*, *15*(4), 537–550. doi:10.1016/j.leaqua.2004.05.006

Downing, K., Chan, S. W., Downing, W. K., Kwong, T., & Lam, T. F. (2008). Measuring gender differences in cognitive functioning. *Multicultural Education & Technology Journal*, *2*(1), 4–18. doi:10.1108/17504970810867124

Dutton, J. E., & Ashford, S. J. (1993). Selling issues to top management. *Academy of Management Review*, *18*(3), 397–428.

Dutton, J. E., Ashford, S. J., O'Neill, R. M., & Lawrence, K. A. (2001). Moves that matter: Issue selling and organizational change. *Academy of Management Journal*, *44*(4), 716–736. doi:10.2307/3069412

Eglene, O., Dawes, S. S., & Schneider, C. A. (2007). Authority and leadership patterns in public sector knowledge networks. *American Review of Public Administration*, *37*(1), 91–113. doi:10.1177/0275074006290799

Eyal, O., & Kark, R. (2004). How do transformational leaders transform organizations? A study of the relationship between leadership and entrepreneurship. *Leadership and Policy in Schools*, *3*(3), 211–235. doi:10.1080/15700760490503715

Ferris, G. R., Treadway, D. C., Kolodinsky, R. W., Hochwarter, W. A., Kacmar, C. J., Douglas, C., & Frink, D. D. (2005). Development and validation of the political skill inventory. *Journal of Management*, *31*(1), 126–152. doi:10.1177/0149206304271386

Frese, M., & Fay, D. (2001). 4. Personal initiative: An active performance concept for work in the 21st century. *Research in Organizational Behavior*, *23*, 133–187. doi:10.1016/S0191-3085(01)23005-6

Frese, M., Fay, D., Hilburger, T., Leng, K., & Tag, A. (1997). The concept of personal initiative: Operationalization, reliability and validity in two German samples. *Journal of Occupational and Organizational Psychology*, *70*(2), 139–161. doi:10.1111/j.2044-8325.1997.tb00639.x

Friedman, M. (1997). Public schools: Make them private. *Education Economics*, *5*(3), 341–344. doi:10.1080/09645299700000026

Giles, C., & Hargreaves, A. (2006). The Sustainability of Innovative Schools as Learning Organizations and Professional Learning Communities During Standardized Reform. *Educational Administration Quarterly*, *42*(1), 124–156. doi:10.1177/0013161X05278189

Grant, A. M., & Ashford, S. J. (2008). The dynamics of proactivity at work. *Research in Organizational Behavior*, *28*, 3–34. doi:10.1016/j.riob.2008.04.002

Grant, A. M., & Parker, S. K. (2009). 7 Redesigning Work Design Theories: The Rise of Relational and Proactive Perspectives. *The Academy of Management Annals*, *3*(1), 317–375. doi:10.1080/19416520903047327

Gunn, L. (1988). Public management: A third approach? *Public Money & Management*, *8*(1-2), 21–25. doi:10.1080/09540968809387457

Gupton, S. L. (1998). *State-Mandated Reform of Programs Preparing School Principals: One University's Experience*. Paper presented at the Annual Meeting of the Mid-South Educational Research Association.

Hambrick, D. C., Cho, T. S., & Chen, M. J. (1996). The influence of top management team heterogeneity on firms' competitive moves. *Administrative Science Quarterly*, *41*(4), 659–684. doi:10.2307/2393871

Hassan, S., & Rohrbaugh, J. (2009). Incongruity in 360-degree feedback ratings and competing managerial values: Evidence from a public agency setting. *International Public Management Journal*, *12*(4), 421–449. doi:10.1080/10967490903103375

Heilbrunn, S. (2010). Advancing entrepreneurship in an elementary school: A case study. *International Education Studies*, *3*(2), 174. doi:10.5539/ies.v3n2p174

Hendrick, R. (2003). Strategic planning environment, process, and performance in public agencies: A comparative study of departments in Milwaukee. *Journal of Public Administration: Research and Theory*, *13*(4), 491–519. doi:10.1093/jopart/mug031

Hooijberg, R., & Choi, J. (2001). The impact of organizational characteristics on leadership effectiveness models an examination of leadership in a private and a public sector organization. *Administration & Society*, *33*(4), 403–431. doi:10.1177/0095399701334002

Howard-Grenville, J. A. (2007). Developing issue-selling effectiveness over time: Issue selling as resourcing. *Organization Science*, *18*(4), 560–577. doi:10.1287/orsc.1070.0266

Juenke, E. G. (2005). Management tenure and network time: How experience affects bureaucratic dynamics. *Journal of Public Administration: Research and Theory*, *15*(1), 113–131. doi:10.1093/jopart/mui006

Keck, S. L., & Tushman, M. L. (1993). Environmental and organizational context and executive team structure. *Academy of Management Journal*, *36*(6), 1314–1344. doi:10.2307/256813

Kimberly, J. R., & Evanisko, M. J. (1981). Organizational innovation: The influence of individual, organizational, and contextual factors on hospital adoption of technological and administrative innovations. *Academy of Management Journal*, *24*(4), 689–713. doi:10.2307/256170 PMID:10253688

LePine, J. A., & Van Dyne, L. (1998). Predicting voice behavior in work groups. *The Journal of Applied Psychology*, *83*(6), 853–868. doi:10.1037/0021-9010.83.6.853

Li, H., & Zhang, Y. (2007). The role of managers' political networking and functional experience in new venture performance: Evidence from China's transition economy. *Strategic Management Journal*, *28*(8), 791–804. doi:10.1002/smj.605

London, M., Larsen, H. H., & Thisted, L. N. (1999). Relationships between feedback and self-development. *Group & Organization Management*, *24*(1), 5–27. doi:10.1177/1059601199241002

Luo, Y. (2003). Industrial dynamics and managerial networking in an emerging market: The case of China. *Strategic Management Journal*, *24*(13), 1315–1327. doi:10.1002/smj.363

Major, D. A., & Kozlowski, S. W. (1997). Newcomer information seeking: Individual and contextual influences. *International Journal of Selection and Assessment*, *5*(1), 16–28. doi:10.1111/1468-2389.00042

Maranto, R. (2003). Lobbying in Disguise. *Education Next*, *3*(1), 79–84.

McCallum, S. Y. L., Forret, M., & Wolff, H. G. (2014). Internal and external networking behavior: An investigation of relationships with affective, continuance, and normative commitment. *Career Development International*, *19*(5), 595–614. doi:10.1108/CDI-08-2013-0101

Miller, D., Lant, T. K., Milliken, F. J., & Kom, H. J. (1996). The evolution of strategic simplicity: Exploring two models of organizational adaption. *Journal of Management*, *22*(6), 863–887. doi:10.1177/014920639602200604

Miller, J. P. (1994). The relationship between organizational culture and environmental scanning: A case study. *Library Trends*, *43*(2), 170–205.

Miller, V. D., & Jablin, F. M. (1991). Information seeking during organizational entry: Influences, tactics, and a model of the process. *Academy of Management Review*, *16*(1), 92–120.

Morrison, E. W. (1993). Longitudinal study of the effects of information seeking on newcomer socialization. *The Journal of Applied Psychology*, *78*(2), 173–183. doi:10.1037/0021-9010.78.2.173

Morrison, E. W., & Phelps, C. C. (1999). Taking charge at work: Extrarole efforts to initiate workplace change. *Academy of Management Journal*, *42*(4), 403–419. doi:10.2307/257011

Mulford, B. (2003). *School leaders: Changing roles and impact on teacher and school effectiveness.* A paper commissioned by the Education and Training Policy Division, OECD, for the Activity Attracting, Developing and Retaining Effective Teachers.

Newman, J., Raine, J., & Skelcher, C. (2000). *Innovation and best practice in local government: A research report.* Department of the Environment, Transport and the Regions.

O'Neil, J. (2004). Effects of gender and power on PR managers' upward influence. *Journal of Managerial Issues*, 127–144.

Othman, R., & Hamedon, S. R. (2011). *Environmental scanning practice of enterprise 50 Small Medium Enterprise (E50 SMEs) in Malaysia.* Retrieved from irep.iium.edu.my

Parker, S. K. (1998). Enhancing role breadth self-efficacy: The roles of job enrichment and other organizational interventions. *The Journal of Applied Psychology*, *83*(6), 835–852. doi:10.1037/0021-9010.83.6.835 PMID:9885197

Parker, S. K., & Collins, C. G. (2010). Taking stock: Integrating and differentiating multiple proactive behaviors. *Journal of Management*, *36*(3), 633–662. doi:10.1177/0149206308321554

Pelled, L. H. (1996). Demographic diversity, conflict, and work group outcomes: An intervening process theory. *Organization Science*, *7*(6), 615–631. doi:10.1287/orsc.7.6.615

Perry, J. L., & Rainey, H. G. (1988). The public-private distinction in organization theory: A critique and research strategy. *Academy of Management Review*, *13*(2), 182–201.

Preble, J. F. (1992). Towards a Comprehensive System Of Strategic Control. *Journal of Management Studies*, *29*(4), 391–408. doi:10.1111/j.1467-6486.1992.tb00671.x

Rainey, H. G. (1983). Public Agencies and Private Firms Incentive Structures, Goals, and Individual Roles. *Administration & Society*, *15*(2), 207–242. doi:10.1177/009539978301500203

Rashman, L., Withers, E., & Hartley, J. (2009). Organizational learning and knowledge in public service organizations: A systematic review of the literature. *International Journal of Management Reviews*, *11*(4), 463–494. doi:10.1111/j.1468-2370.2009.00257.x

Reichers, A. E. (1987). An Interactionist Perspective on Newcomer Socialization Rates. *Academy of Management Review*, *12*(2), 278–287.

Renn, R. W., & Fedor, D. B. (2001). Development and field test of a feedback seeking, self-efficacy, and goal setting model of work performance. *Journal of Management*, *27*(5), 563–583. doi:10.1177/014920630102700504

Russell, R. D., & Russell, C. J. (1992). An Examination of the Effects of Organizational Norms, Organizational Structure, Environmental Uncertainty on Entrepreneurial Strategy. *Journal of Management*, *18*(4), 639–656. doi:10.1177/014920639201800403

Salanova, M., & Schaufeli, W. B. (2008). A cross-national study of work engagement as a mediator between job resources and proactive behaviour. *International Journal of Human Resource Management*, *19*(1), 116–131. doi:10.1080/09585190701763982

Sass, C. P. (1994). *On Interpersonal Competence. Interpretive Approaches to Interpersonal Communication* (K. Carter & M. Presnell, Eds.). Albany, NY: State University of New York Press.

Scott, S. G., & Bruce, R. A. (1994). Determinants of innovative behavior: A path model of individual innovation in the workplace. *Academy of Management Journal*, *37*(3), 580–607. doi:10.2307/256701

Şeker, M. (2011). *İstanbul'da Yaşam Kalitesi Araştırması*. İstanbul: İstanbul Ticaret Odası yayınları.

Škerlavaj, M., Štemberger, M. I., & Dimovski, V. (2007). Organizational learning culture—the missing link between business process change and organizational performance. *International Journal of Production Economics*, *106*(2), 346–367. doi:10.1016/j.ijpe.2006.07.009

Skule, S. (2004). Learning conditions at work: A framework to understand and assess informal learning in the workplace. *International Journal of Training and Development*, *8*(1), 8–20. doi:10.1111/j.1360-3736.2004.00192.x

Sonfield, M., Lussier, R., Corman, J., & McKinney, M. (2001). Gender comparisons in strategic decision-making: An empirical analysis of the entrepreneurial strategy matrix. *Journal of Small Business Management*, *39*(2), 165–173. doi:10.1111/1540-627X.00015

Teunissen, P. W., Stapel, D. A., van Der Vleuten, C., Scherpbier, A., Boor, K., & Scheele, F. (2009). Who wants feedback? An investigation of the variables influencing residents' feedback-seeking behavior in relation to night shifts. *Academic Medicine*, *84*(7), 910–917. doi:10.1097/ACM.0b013e3181a858ad PMID:19550188

Thomas, J. B., Clark, S. M., & Gioia, D. A. (1993). Strategic sensemaking and organizational performance: Linkages among scanning, interpretation, action, and outcomes. *Academy of Management Journal*, *36*(2), 239–270. doi:10.2307/256522 PMID:10125120

Thomas, J. B., Sussman, S. W., & Henderson, J. C. (2001). Understanding "strategic learning": Linking organizational learning, knowledge management, and sensemaking. *Organization Science*, *12*(3), 331–345. doi:10.1287/orsc.12.3.331.10105

Thompson, J. A. (2005). Proactive personality and job performance: A social capital perspective. *The Journal of Applied Psychology*, *90*(5), 1011–1017. doi:10.1037/0021-9010.90.5.1011 PMID:16162073

Thompson, S. C., Gregg, L., & Niska, J. M. (2004). Professional Learning Communities, Leadership, and Student Learning. *RMLE Online: Research in Middle Level Education*, *28*(1), 1–15. doi:10.1080/19404476.2004.11658173

Todd, S. Y., Harris, K. J., Harris, R. B., & Wheeler, A. R. (2009). Career success implications of political skill. *The Journal of Social Psychology*, *149*(3), 279–304. doi:10.3200/SOCP.149.3.279-304 PMID:19537596

VandeWalle, D., Ganesan, S., Challagalla, G. N., & Brown, S. P. (2000). An integrated model of feedback-seeking behavior: Disposition, context, and cognition. *The Journal of Applied Psychology*, *85*(6), 996–1003. doi:10.1037/0021-9010.85.6.996 PMID:11125662

Wanberg, C. R., & Kammeyer-Mueller, J. D. (2000). Predictors and outcomes of proactivity in the socialization process. *The Journal of Applied Psychology*, *85*(3), 373–385. doi:10.1037/0021-9010.85.3.373 PMID:10900812

Wang, J., Tolson, H., Chiang, T. L., & Huang, T. Y. (2010). An exploratory factor analysis of workplace learning, job satisfaction, and organizational commitment in small to midsize enterprises in Taiwan. *Human Resource Development International*, *13*(2), 147–163. doi:10.1080/13678861003703682

West, M. A., & Anderson, N. R. (1996). Innovation in top management teams. *The Journal of Applied Psychology*, *81*(6), 680–693. doi:10.1037/0021-9010.81.6.680

Yuan, F., & Woodman, R. W. (2010). Innovative behavior in the workplace: The role of performance and image outcome expectations. *Academy of Management Journal*, *53*(2), 323–342. doi:10.5465/AMJ.2010.49388995

ADDITIONAL READING

Balcı, A. (2014). Managing on the Edge of Chaos. In Chaos, Complexity and Leadership 2012 (pp. 119-129). Springer Netherlands.

Grant, A. M., Gino, F., & Hofmann, D. A. (2011). Reversing the extraverted leadership advantage: The role of employee proactivity. *Academy of Management Journal*, *54*(3), 528–550. doi:10.5465/AMJ.2011.61968043

Parker, S. K., Bindl, U. K., & Strauss, K. (2010). Making things happen: A model of proactive motivation. *Journal of Management*, *36*(4), 827–856. doi:10.1177/0149206310363732

KEY TERMS AND DEFINITIONS

Feedback Seeking: Efforts undertaken to determine the accuracy and sufficiency of behaviors displayed to reach the work goals.

Innovativeness: Implementation of new ideas, products, processes and procedures in organizations. It is either the creation of new constructive changes or the improvement of the existing resources to enhance organizational effectiveness.

Issue Selling: A behavior displayed by employees to draw the attention of other members of the organization and to influence their understanding of the events, developments, and trends that have effects on organizational performance.

Proactivity: An anticipatory action that employees take to create or control a situation. Employees need to take initiative, anticipate events, take measures and act in advance in order to fulfil a proactive behaviour.

Social Networking: A behavior displayed by employees to build a social environment which may help them to be successful in the workplace.

Strategic Learning: An information seeking and acquistion activity performed to prepare for potential future problems or opportunities.

Strategic Scanning: An information acquisition activity carried out for exploratory purposes. The information gathered is used to identify the potential problems and opportunities before they ocur.

APPENDIX

Table 10. Proactive manager scale

Item	Factor Loadings	Item Total Corr.
Innovativeness Cronbach Alpha = .829 Variance explained = 15.02		
1. I take the lead in implementing new methods and techniques which may improve the effectiveness of my organization.	0.789	0.695
2. I overcome obstacles at work by trying different and new methods.	0.697	0.556
3. I take the initiative to solve problems at work.	0.673	0.576
4. I make constructive suggestions to my co-workers for performing their tasks more effectively.	0.642	0.630
5. I try to change the rules and processes hindering the effective execution of tasks within my organization.	0.640	0.596
6. I try to ensure that my suggestions for performing the work processes more effectively are followed.	0.631	0.586
Issue Selling Cronbach Alpha = .752 Variance explained =10.53%		
7. I take a leading role in discussions about the running of my organization.	0.752	0.580
8. I try to get the solutions I propose for the problems of my organization taken into consideration.	0.732	0.498
9. I try to ensure that my superiors take precautions against potential organizational crises.	0.723	0.559
10. I mobilize my superiors and co-workes to enable my organization to take advantage of potential opportunities.	0.679	0.573
Social Networking Cronbach Alpha = .789 Variance explained = 10.21%		
11. I try to meet people who may be able to support me in realizing my future career objectives.	0.870	0.746
12. I participate in social activities to develop new working relationships.	0.767	0.644
13. I mention my career plans to influential people (i.e. politician, unionist, top manager, etc.) to gain their support.	0.723	0.551
14. I try to make close friendships with people whom I can benefit from their knowledge.	0.624	0.482
Strategic Scanning Cronbach Alpha = .804 Variance explained = 9.85%		
15. I try to anticipate the potential problems to take necessary precautions before they occur.	0.655	0.585
16. I try to identify the potential opportunities to include them in my plans.	0.653	0.624
17. I take the problems and opportunities I may encounter into consideration while determining my objectives.	0.611	0.565
18. I carry out analyses to identify the problems I may encounter.	0.494	0.594
19. I prepare plans to take advantage of the opportunities I may encounter.	0.471	0.600

continued on following page

Table 10. Continued

Item	Factor Loadings	Item Total Corr.
Strategic Learning Cronbach Alpha = .978 Variance explained = 8.45%		
20. Before starting a task, I collect information about the required qualifications and performance criteria.	0.710	0.458
21. I take advantage of learning opportunities which provide an improvement in my work performance.	0.682	0.512
22. I turn complex and difficult issues I encounter at work into learning opportunities.	0.679	0.439
23. I carefully review written resources (guidelines, job descriptions, regulations, etc.) regarding the content of my task.	0.655	0.437
Feedback Seeking Cronbach Alpha = .594 Variance explained = 6.80%		
24. I share my work plans with my superiors and co-workers to get their feedback.	0.798	0.535
25. I hold meetings periodically to obtain feedback from my co-workers about my performance.	0.699	0.412
26. I carefully observe the reactions of my co-workers to form an opinion about my own performance.	0.351	0.301

Cronbach Alpha= .88
Overall Variance Explained = 60.85%

Chapter 4
Leaders' Emotional Intelligence in the Chaos and Complexity Situations of Educational Organizations

Şefika Şule Erçetin
Hacettepe University, Turkey

Barış Eriçok
Nevşehir Hacı Bektaş Veli University, Turkey

Anıl Kadir Eranıl
Nevşehir Provincial Directorate of National Education, Turkey

ABSTRACT

This study examines that in what aspects the levels of leaders' emotional intelligence in chaos and complexity situations can affect the organizations. Also, in this study the dimensions of emotional intelligence which are closely related to social life and organizational systems will be considered in terms of chaos and complexity theories. At the same time in this study, how a situation that appears to be complex but has an order inside and has the potential to be directed with small interventions can be handled by a leader that has a high capacity of using his/her emotional intelligence will be scrutinised. Additionally, this study will present the advantages of some leader capabilities (such as personal and interpersonal relationships, stress

DOI: 10.4018/978-1-5225-0460-3.ch004

management, adaptability, general mood, empathy, motivation and self-awareness) in the management of the organizations in the context of chaos and complexity theories. Last but not least, the effect of emotional intelligence level in the rising of new leaders will be mentioned in this study.

INTRODUCTION

Educational organizations involve dynamic systems. The dynamic systems that educational organizations involve contain a lot of closely interrelated sub-systems inside. In these sub-systems, educational institutions serve to human beings with other human beings. For this reason, the relationships of the sub-systems of educational institutions are different from the traditional organizations and they have complex and multi-perspective order. Unlike the traditional organization systems, the order in chaotic systems depends on some basic principles such as non-linearity, iterations and sensitivity to initial conditions. It can be asserted that educational institutions have chaotic organization characteristics because the relationships among the sub-systems of educational organizations cannot be handled in a linear way, there are several cycles (iterations) in educational organizations and small changes in educational organizations can cause big outcomes. It can be said that the chaotic structures of educational institutions requires leaders that can pioneer the change with small interventions. The control of the leader on the whole organization should work in a different way from the traditional leadership perception. In this sense, it can be thought that the focus point of the leadership concept has changed from simply effecting people into initiating the relationships that will drive the organization to the aimed level by means of small interventions in the context of order, disorder and change. When it comes to emotional intelligence, it is a concept that includes understanding others and controlling emotions by taking self-awareness as its starting point. In this study, analysing the reflections of leaders' levels of using their emotional intelligence in chaos and complexity situations is aimed. In this context, the levels and styles of the leaders' emotional intelligence can affect the future of their own school in the first step and the whole educational system in the next phase.

THEORIES OF CHAOS AND COMPLEXITY

Chaos is everywhere and anywhere. It sometimes seems to be unpredictable or vague but it is possible to see, evaluate and perceive its cues and manage the chaos with intelligence of the human factor and the organizations (Kayman, 2014, p. 289). Accordingly, it can be thought that chaos and complexity theories seek answer to several questions in the world of science. Is the obscurity in the universe a continuation of a pattern system? If the universe has a pattern system in itself, do all the units that form the universe have a pattern system inside? Do these systems consist of a set of order and disorder? Is the disorder an order that exists in the big order?

Lorenz (1993) defines chaos as a process that "appears to proceed according to chance even though their behaviour is in fact determined by precise laws". According to Tetenbaum (1998, p. 24) chaos is a complex, unpredictable, and orderly disorder in which patterns of behaviour unfold in irregular but similar forms. The regular irregularity of snowflakes that have an always-recognizable six-sided figure, but each snowflake unique can be seen as an example. In addition to this, Koçak (2000, p. 94) explains chaos on the basis of the movements of a floating leaf:

When a leaf is drawn into a vortex during its movement, it takes a tour around the vortex. When the leaf gets rid of the vortex, it continues the journey until it is drawn into a new vortex. A very small change in the position of the leaf changes over all its behaviours in the future. That the principal of small changes can make a big difference is one of the distinct features of chaos theory. Chaos can be seen anywhere in nature, sometimes in a dripping tap or sometimes in the heart-beats of human.

According to Altunoğlu (2009, p. 8) even if chaos theory hasn't completed its evolution yet, it is a new and comprehensive point of view with its geometry, fuzzy logic in computer world, dynamic systems in physic, neural analysis in biology and new models in economics. Along similar lines, Erçetin, Bisaso and Saeed (2013, p. 149) argues that

Chaos Theory represents a big jump from the way we have thought in the past— a paradigm shift. The traditional notion of chaos is simply; unorganized, disorderly, random etc. But Chaos theory has nothing do with the traditional notion per se. On the contrary, it actually tells you that not all that 'chaos' you see is due to chance, or random or caused by unknown factors. Chaos theory is thus about the deterministic factors (non-linear relationships) that cause things to look random.

Ferreira, Filipe, Coelho and Pedro (2014, p. 89) states that relationships in non-linear systems are unstable. Chaos theory aims to understand and explain this kind of unpredictable aspects of nature, social life, the uncertainties, the nonlinearities, the disorders and confusion. Similarly, according to Ural and Açıkalın (2015, p. 200) "chaos theory deals with the changing relationship between order and disorder in the behavior of natural or social systems". From the definitions that have been mentioned above, it is possible to deduce that chaos theory can be transferred into the most parts of life and it includes a relationship of an order and disorder in itself.

Based on the ideas of Tetenbaum (1998, p. 21), complexity theory is a spin-off version of chaos theory. However, McMillan (2004, p. 25) states that because of being a new and broad theory, it is difficult to make a clear definition and set the boundaries of complexity theory. It is an interdisciplinary area and this situation has been growing by means of interdisciplinary applications. Allen (2001, p. 150) also maintains that complex systems have capacity to respond their environment sophisticatedly. This means that they are not mechanical systems with a single way, but they have some internal possibilities of choice or responses. In the same way, Marion (2006) hypothesises that complexity theory is a theory that emerged in the last twenty years with the aim of exploring the nature. To sum up, it can be said that complexity theory has evolved out of chaos theory and it includes non-linear contacts in itself.

Chaos and complexity theory can be exemplified through a scaffolding sample. Scaffoldings, which are used in the building process of the structures and surround them, can be seen as a disorganized iron pile at first glance. On the other hand, scaffoldings contain an order within themselves. Every individual part of the scaffoldings has a significant role on the whole. If one or some parts of the scaffolding are positioned in an inappropriate way, the collapse of the whole can accelerate. From this point of view, the systems that appear to be complex but have an order in themselves can be explained by chaos and complexity theories. As before mentioned, complex looking systems can be seen in organizations because there are lots of interconnected sub-units in organizations. Even if the relationship between these sub-units and other parts of the organizations seems complex, it follows an order. Along similar lines, according to Tetenbaum (1998, p. 21), chaos theory and complexity theory argue that relationships in complex systems are nonlinear. They are composed of interconnected and multiple alternatives and these alternatives can create unpredictable results. All in all, the reflections of chaos theory can be seen in organizations because the unexpected events and unplanned consequences can arise in organizations. These events and consequences can both create an order or they can cause a disorder.

Educational institutions that held an extensive place in organizational systems can be handled within the scope of chaos theory because educational institutions include some kind of disorder and non-linear structures in themselves. As Mason (2008, p. 19) says, educational organizations exhibit many features of complex adaptive systems because of being dynamical and unpredictable, being non-linear organizations operating in unpredictable and changing external environments. Correspondingly, Açıkalın and Bölücek (2014, p. 33) states that "chaos can occur in nonlinear and dynamic systems". It has been shown that, examining the educational organizations in the scope of chaos theory is possible because educational organizations have some characteristics of chaos theory like being nonlinear and dynamic.

CHAOS, COMPLEXITY, AND LEADERSHIP IN EDUCATIONAL ORGANIZATIONS

Şişman (1994, p. 29) defines educational organizations as the organizations whose input and output is human, in other words, the organizations in which some people produce service for other people. Accordingly, the individual and the society underlie the educational organizations (Özdemir & Kılınç, 2014, p. 76). From the above, it can be thought that educational organizations have a role in shaping the societies.

As a social organization, the school has two sides, one is formal side and the other is informal side. The formal side of school contains the structural characteristics of the school. On the contrary, the informal side of school involves the individual characteristics of the school (Bursalıoğlu, 2013, p. 135). The individual characteristics of the educational organizations (informal side) may be much more sensitive than their structural characteristics (formal side). For this reason, "educational organizations are unique and seeing them only as bureaucratic organizations means ignoring the informal side of the schools that prevail the formal side" (Uğurlu, 2015, p. 211). The structure and functioning of the educational organizations can be said to be the formal side of schools. In this context, organization charts of schools represent the formal side of schools because these organization charts consist of law-regulated superior-subordinate relationships. On the other hand, the human relations aspect of educational organizations, which is unpredictable and not regulated by law, represents the informal side of educational organizations. Furthermore, the everyday problems of educational organizations that are unpredictable and can be overcame only by individual and social relationships can also be held in terms of the informal side of education. The formal side of the educational organizations can be said to be the visible part of the iceberg and the informal side to be the invisible part because while the matters related to the formal side of the educational organizations can be predicted easily, those related to the informal side cannot.

Organizational systems can be discussed under two headings: mechanistic and organic models. Mechanistic systems (bureaucracies) define functions, responsibilities and powers in other words it can be said that boundaries are set. Mechanistic system dictate people where and how to behave, which behaviour is appropriate and which is not or which behaviour is expected and which is not. When it comes to the organic systems, they are not hierarchic in the same sense as are mechanistic. Organic systems are flexible to fit the changing situations. Positions are differentiated according to seniority or expertise in organic systems. The authority is often the most capable and informed one and the location of authority is settled by consensus. Finally, the extent to which the individual yields himself as a resource to be used by the working organization is far more extensive in organic than in mechanistic systems (Burns & Stalker 1961, pp. 103-108). The mechanistic and organic models can be examined below.

According to the Figure 1, it can be said that mechanistic systems include a structure that is connected to the head and follow hierarchical top-down design. On the other hand, organic systems follow a more complicated order and they have cross-functional teams. It is apparent that there is a similarity between the formal side of educational organizations and mechanistic model of organization. Formal side of schools and the mechanistic organizational structure follow a linear order with Newtonian perspective. Complementary to this, it can be said to be a similarity between the informal side of educational institutions and organic model of organization. Informal side of educational institutions and organic model of organizations involve many unpredicted and chaotic situations. In these situations there is a nonlinear order. Additionally, Erçetin, Bisaso and Saeed (2015, p. 159) argued that "education is a dynamic and complex system that evolves to meet the demands of society." It can be seen that educational institutions evolve and update themselves in order to meet the needs of the society.

Chaos theory is very appealing to the educators because in the everyday routines of the educational institutions, some situations can arise contrary to the expectations and some chaotic situations can emerge as a result. It can be thought that the objectives of the educational institutions can be achieved but this process rarely progresses in a predictable way. Chaotic situations can always be seen in educational organizations. These situations can be explained and illustrated with the following examples: social breakdown of communities, need to reassess grading system, passing kids who are not ready, less student accountability, too little money, not enough structure for students, parent influence overshadows student needs, schools are social centres rather than learning organizations, schools haven't changed, but students have changed, dysfunctional families, low parental involvement, student's lack

Figure 1. The mechanistic and organic models
Source: Robbins, S. P., & Judge, T. A. (2003). Organizational Behaviour. New Jersey: Pearson Education.

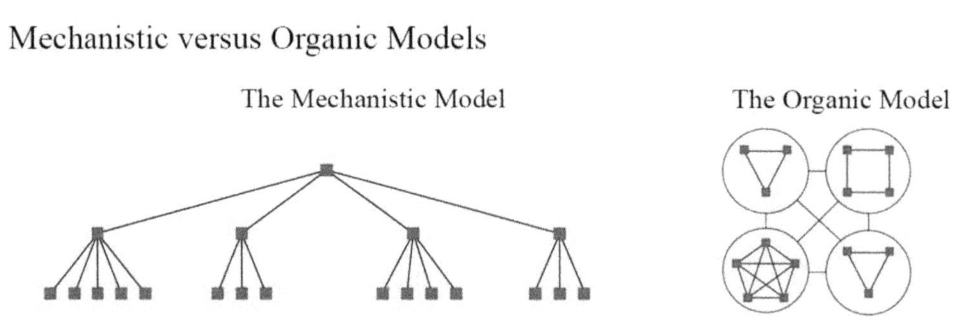

of motivation, overcrowded classes, bureaucratic policies and emphasis is placed on paperwork, not on teaching (Snyder, Acker-Hocevar & Wolf, 1995, pp. 7-12). To sum up, the reason of chaotic situations are often seen in educational organizations is due to the fact that educational institutions are organic organizations that produce service by people to other people. As is seen, a limited number of predictions and distributed roles determined by the formal side of educational institutions are not enough to estimate the unlimited number of chaotic situations that are possible to happen in the routine of educational institutions.

Around 500 B.C, the Greek philosopher Heraclitus addressed the idea that the universe is in a constant state of flux and it has the characteristics of both permanence and change by saying "you cannot step twice into the same river, for other waters are continually flowing on". As he noted, "everything flows and nothing abides; everything gives way and noting stays fixed… Cool things become warm, the warm grows cool; the moist dries, the parched becomes moist." Heraclitus states that the secrets of the universe is in the hidden connections that constitutes the unity and change at the same time (Morgan, 1998, p. 240). The change extending from the depth of the universe to the social structures, according to Schlechty (2011, p. 40) can only be maintained by means of a leader who acts as a change agent and a system that support the change.

Morgan (1998, pp. 255-263) sorted five principles to guide the change management of organizations with the perspective of chaos and complexity theories. The first one is *rethinking organization* and according to this principle managers should rethink the nature of order and organization. It should be known that the hierarchic order and predetermined logic that works according to the traditional organization rules will be replaced by the new order because new order emerges in all organizations that is pushed into the edge of chaos by internal and external fluctuations. This new order is natural, emergent and free but can never be planned or predetermined. The second principle is *the art of managing and changing context* and it is about shaping and creating contexts in which self-organization can occur. According to this principle managers should produce new understandings and take part in actions to create new contexts. Otherwise, the system can be locked and no change can be achieved. The third principle, *using small changes to create large effects,* includes the idea that small but critical changes at critical times can trigger crucial effects. The fourth principle is *living with emergence as a natural state of affairs.* No one has ability to control or predetermine the incidents in complex systems. New situations emerge and develop. The thing that the manager should do is to nudge and push the system in a desired direction by means of critical but effective interventions. The fifth and last principle is being to open to new metaphors that can facilitate self-organization. This principle is about being open to some metaphors that is successful in terms of self-organizing like termite colonies and beehives. With regard to the examination of the organization in terms of chaos and complexity theories, Erçetin, Potas, Kısa and Açıkalın (2013, p. 189) argues that "when complexity is recognized in a system's environment, it must be taken into consideration as soon as possible". To conclude, a leader should manage and care the chaotic situations in an effective way and swap the centralised organization understanding with the multi-centred (decentralized) organization because as it is mentioned above the leader has a crucial role in managing the change with chaos and complexity perspectives.

Bass (2009, p. 25) interpreted leadership as an interaction between members of a group that often involves a structuring or restructuring of the situation and of the perceptions and expectations of the members. Equally important, leaders are agents of change, whose acts affect others more than other people's acts affect them. Another outstanding definition of leadership is that leadership is "influencing others to come together around a common vision (Daft, 2014, p. 5)". As it can be understood from the key aspects discussed above, leaders have multiple roles and these roles require leaders to act in nonlinear situations because leaders are in the position of managing both the structural aspects of their organizations and the complex human relations of the members. To sum, in educational organizations human beings serve to other human beings and this unique characteristic of educational organizations requires the need to leaders that have the ability to manage complex relations.

Bursalıoğlu (2013, p. 204) states that according to Cartwright-Zander (1958) leader is "someone who creates a considerable effect on a group and the measure of the leadership is the change based on this effect". According to Follet (1940) "leader is the one who interprets the experiences of the group, change them towards the goals of the organization and so benefit from the energy of the group". In this context, considering the leader definitions above, it can be concluded that leaders are required to have some characteristics such as vision, creativity, holding the group together and problem solving. Moreover, it can be thought that leadership is an impact and guidance oriented concept. When the organizational structure of the educational institutions is considered, the leadership concept has a crucial role because managing changeable and lively nature of the educational organizations is possible through a leader.

The predictable and determinable actions in the Newtonian perspective leave their place to quantum mechanics. According to quantum mechanics and chaos theory, there is a new understanding of order, disorder and change. In quantum mechanics, individual actions of atoms or people cannot be easily predicted and controlled. Nothing can exist independently from its relationships with others, so the relationships are the key determinants of a well-ordered system. The order emerges not as a result of controlling individual parts, instead it emerges via the relationships that make up the whole system. Like all open systems, organizations grow and change as a result of disequilibrium and disorder. The disorder turns into a new order.

These new understandings provide some implications for leaders. These implications can include supporting relationships with a clear vision, information sharing, expressing of values and giving up the strict rules and controls. Moreover, the new understandings make the leaders focus on the whole, not on the parts. Finally, leaders experience a change in reducing the boundaries between departments and organizations to allow new patterns of relationships (Daft, 2014, p. 65; Wheatley, 2010). To review, it can be said that the new understandings of science change the structures of organizations and the new structures of organizations change the leaders. As an example, Erçetin, Açıkalın and Bülbül (2012, pp. 89-104) have developed a new leadership type within the frame of chaos and complexity theory. According to Erçetin (2015, p. 6), this new type is plasma leadership and it requires coping with chaotic situations.

As a result of the new developments and implications of science, it can be concluded that the focus point of leadership has shifted from simply effecting people to initiating the relationships that push the organization to the desired point within the context of order, disorder and change with small but effective interventions. In initiating these relationships, the behaviour types of the leaders are very influential because leaders have a crucial role in canalising the emotions of the members into the favour of the organizations.

THE EMOTIONAL INTELLIGENCE OF LEADERS IN CHAOS AND COMPLEXITY

The fine line between being successful only in school or in social life can be explained by the using level of emotional intelligence. While academic success means being successful only in an exam or in a course, social success is much more of it. Being socially successful means having good relationships, having ability to analyse oneself and the environment, managing the complex inner circle relationships, being resistant to difficulties and being successful at the workplace. Therefore, level of emotional intelligence, different from the intelligence quotient, can be seen as a determiner of total success in life.

Goleman (2013, p. 393) defines emotional intelligence as recognizing one's own emotions and emotions of others, motivating oneself and managing the relationships well. Goleman model of emotional intelligence is composed of two parts: personal competences and social competences. Personal competences include self-awareness, self-regulation and motivation and social competences include empathy and social skills. Moreover, there are twenty-four other emotional competences under these headings.

According to Bar-On (2006, pp. 3-4), emotional intelligence has the ability to recognize, understand and express emotions and feelings and manage them, to adapt and solve problems and to have a positive mood and be self-motivated. In his model, Bar-On identified five competencies and 15 sub-skills. These are intrapersonal competencies (self-regard, emotional self-awareness, assertiveness, independence and self-actualization); interpersonal (empathy, social responsibility and interpersonal relationship); stress management (stress tolerance and impulse control); adaptability (reality-testing, flexibility and problem-solving); and general mood (optimism and happiness).

Stogdill (1948, pp. 63-67) performed a study about leadership and after the detailed examinations of this study he obtained significant findings. Based on these findings, Stogdill lists the topics related to leadership as follow:

1. **Capacity:** Intelligence, alertness, verbal facility, originality, judgement.
2. **Achievement:** Scholarship, knowledge, athletic accomplishments.
3. **Responsibility:** Dependability, initiative, persistence, aggressiveness, self-confidence, desire to excel.
4. **Participation:** Activity, sociability, cooperation, adaptability, humour.
5. **Status:** Socio-economic position, popularity.
6. **Situation:** Mental level, status, skills, needs and interests of followers, objectives to be achieved.

As the above study shows, it can be argued that leadership requires having many characteristics simultaneously. More importantly, it can be argued that emotional intelligence is necessary in the rising of new leaders because it is possible to see the traces of emotional intelligence behind the leader characteristics listed above. Emotional intelligence is a concept that takes one's own emotional awareness to the starting point and by means of this awareness one can understand the other people in the circle. Moreover, emotional intelligence includes some concepts such as empathy, adaptability, optimism, motivation, happiness, self-confidence, practical intuition, interpersonal skills, reliability, desire to success and leadership. When carefully examined, it can be seen that study results of Stogdill (1948, pp. 63-67) and emotional intelligence competencies overlap in general terms. Considering this overlap, it can be concluded that emotional intelligence can contribute to the emergence of leadership skills.

Lucia (1997, p. 25) emphasizes that leadership means empathy. Meanwhile leadership is to know how to listen the other people because big ears and big hearts are common characteristics of effective leaders. Today's chaotic business environment requires leaders to listen their members and members to be heard. When someone ask you to listen him/her, s/he may be begging to be heard by heart. When the leader spare time for him/her and listen with empathy, a success-driven business environment together with trust and commitment emerges. It can be asserted that emotional intelligence lies behind the leadership characteristics explained above because in Goleman's emotional intelligence model (2013, pp. 393-394), which is adapted from Mayer and Salovey, there is "empathy" skill within emotional and social competencies. According to Goleman (2013, pp. 393-394), empathy is the ability to sense the emotions of other people, to put oneself in another person's place and to fit into different environments. As is seen above, it is clear that one of the expected leader characteristics, empathy, has its origins in emotional intelligence. In the work of Lucia above, the leaders' intuition about their members' thoughts is also emphasized. Thus, again, it can be asserted that the leaders' level of emotional intelligence has an undeniable role in making them effective leaders.

As Erçetin, Açıkalın and Bülbül (2012, p. 93) states, "chaos and leadership depend on initial conditions. The way one influences the other develops as in interrelated, non-linear systems". As it is seen, chaos, like leadership, includes a sensitivity to its initial condition. In other words, what results can arise at the end of a leader behaviour cannot be estimated because a strategic or poor attitude that a leader has in a complex situation can change the course of the situation completely. Another essential point is the problematic situations. In these situations the leaders' using level of their emotional intelligence can also have a determining effect because emotional intelligence includes problem solving skills within the dimension of adaptability. For this reason, emotional intelligence level (within the sub-dimension of problem solving) in chaotic and complex situations can play a considerable role in the rising of new leaders and possible results. In parallel, Erçetin and Düzer (2008, pp. 28-30) attributed their own leadership model, Perceptual Multi-dimensional Leadership Model, to five assumptions. Accordingly, (1) everyone is a potential leader. (2) The basis of leadership is made up of perception. (3) The status of leadership is determined by the perception of the leaders and his/her followers. (4) Evaluation of leadership can be carried out of at a specific time point. (5) The perceptual status of leadership cannot carry a judgement whether a leader is successful or not. It can be interpreted that according to the basic assumptions of the leadership model explained above, some emotional intelligence competencies like "empathy (Bar-On, 2006, p. 21)" and "practical intuition (Cooper and Sawaf, 1998, p. xxvii)" can have a notable effect on leadership.

Maddock and Fulton (1998, p. 5) defines leadership as motivation and leader as motivator. When the literature is examined, it can be seen that motivation factor is directly related to emotional intelligence (Goleman, 2013, pp. 393-394; Sengupta, 2009, p. 25; Duncan, 2002. p. 52; Cooper & Sawaf, 1998, p. xxvii). Accordingly, the leaders' using level of emotional intelligence can push them a step further.

According to some people, emotionality is the antithesis of rationality but, in fact, they are interpenetrated. Emotions are often practical for organizations and they are an integral and inseparable parts organizations (Ashforth & Humphrey 1995, p. 97) because people who have complex and nonlinear relationships in their routine transfer these complexities to their organizations. In other words, in every place including organizations the concept of emotion rises. In this context, there are great number of studies that emphasize the emotional side of organizations that some of them can be listed as follow: Sieben and Wettergren, 2010; Zerbe, Härtel and Ashkanasy, 2010; Beyer and Nino, 2001.

In the modern world, the lives of human beings have become more complicated when compared to even a year ago. People have several mental processes and calculate complex possibilities to manage different tasks simultaneously. In a similar manner, science and technology, which is believed to ease the lives of people, engulf people in complex situations by creating new needs. Busy working hours direct people to make time management. For instance, heavy traffic force people to use different alternatives for arriving their destinations. All these complexities manifest themselves in organizations. As the power, size, number of employees and influence circle of the organizations change their structures also become complicated. It is expected from the managers of these complex organizations to have the ability to cope with these complexities because in the changing world and in the new understanding of organization a leader that has the ability of conducting complex relationships is needed.

School managers and all stakeholders ought to know that contemporary education systems are premised on nonlinearity. In a linear system there is a simple cause an effect relationship; A causes B which causes C, and so on. However, a chaotic system is nonlinear. A may not necessarily cause B at all times. Lots of variables come to play and interact with each other. School systems look like nonlinear chaotic systems, too. (Erçetin, Bisaso & Saeed, 2015, p. 154).

In this context, the approach of the educational leaders in managing their relationships with their employees is very crucial. The interaction between an educational leader and a follower don not affect just themselves but the whole educational organization. This evidence shows that educational organizations have a nonlinear structure and the results of an event may not always be estimated. Considering the nonlinearity of the organizations, it can be argued that the emotional intelligence level of the leaders within the context of interpersonal relationships can prevent organizations to mingle with chaotic situations. To sum up, leaders' emotional intelligence levels can come into play in managing the interpersonal relationships.

When the literature is examined, it is seen that the concept of fractal, which is discussed within chaotic organizations, was added into the literature by coining from the adjective of fractus by Mandelbrot (1983, pp. 3-4). According to the dictionary of mathematics, (Clapham and Nicholson, 2009, p. 327), "fractals are typically sets with infinitely complex structure and usually possess some measure of self-similarity". Similarly, "fractals depict chaotic behaviour, yet if one looks closely enough, it is always possible to spot glimpses of self-similarity within a fractal" (Erçetin, Bisaso & Saeed, 2015, p. 152). In this context, fractals can be explained as complex repeating patterns that are self-similar at every scale. Educational institutions are thought to have fractal properties. The similar background of every student at school is an example of fractal. Every student has a family, social circle and some courses at school. On that sense, educational institutions possess small and repeating patterns and they are self-similar at every scale. Every student that joins the educational institution gives life to a new system and in a similar way, every student that quits from the educational institution terminates his/her own system. This circle in which a set of new small systems are integrated into the larger systems and a set of new small systems break away from the larger systems confirms the existence of fractal patterns in educational institutions. The fractal patterns that educational institutions have can be checked out in Figure 2.

When Figure 2 is examined, one can comment on the emotional intelligence levels of managers in the context of the school courses of students which exemplify only one aspect of the student background. The managers who have a role in providing the balance between students' social circles, together with their families,

Figure 2. The fractal patterns in educational institutions

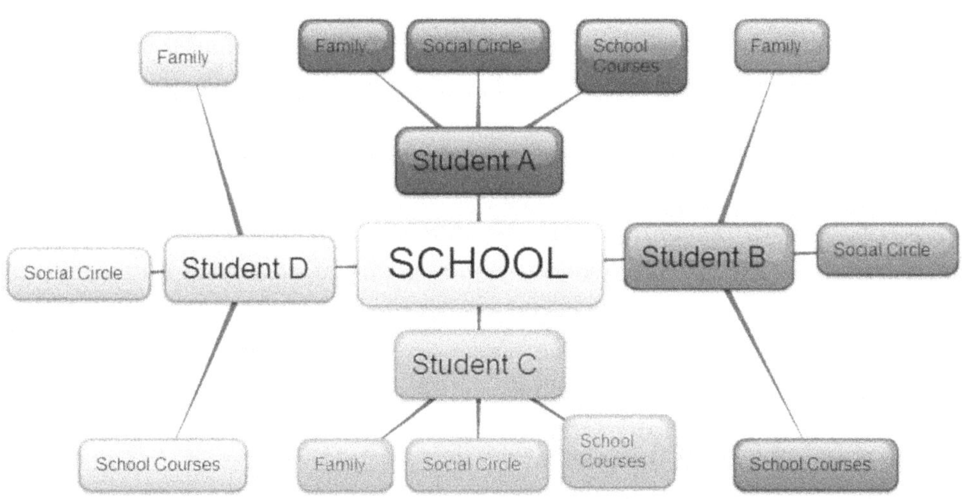

and their schools can act as a source of motivation and empathy in student success in school. In this sense, they can utilize emotional intelligence competencies like motivation and empathy stated by Goleman (2013, pp. 38-39). In the above figure, the family and social circle of students are conspicuous and they're worth mentioning. It can be supposed that every student have to be perceived to have a family and social circle. When it is considered within the context of schools, the relationship of students with their families and social circles needs to be managed gracefully. The balance between the students' family and their social circle is also thought to be one of the concern of school managers. Using of emotional intelligence on the interpersonal level may be necessary when proving the above mentioned balance. The interpersonal level of emotional intelligence includes the skills that enable a person to handle the everyday relationships. Moreover, emotional intelligence on the interpersonal level also includes "empathy, social responsibility and interpersonal relationship skills (Bar-On, 2006)".

In the back of the fractal properties that educational institutions have, there may be some unpredicted complex sub-systems. These complex sub-systems are belong to the students who have different developmental stages at different times, have totally different backgrounds and problems but lead a school life in the same place. At first glance one can think that there is the same sub-system in the background of every student. As an example one can assert that every student has a similar family, social circle and school courses but the whole picture may not be clear every time. The family of each student that s/he has or hasn't may have dreadfully distinguishing features. When looking from the side of educational institution, each student may not have an expected family coherence. Besides, the social circle of a student may not satisfy him/her. Lastly, each student may have different levels of learning at school. While a student may enjoy in a lesson, the other may find it awfully difficult and experience loss of motivation. Considering these factors it can be concluded that educational institutions possess numerous unpredicted complex sub-systems. Similarly, Erçetin (2015, p. 41) describes the complexity of a system as follows:

The complexity of a system in terms of the number of elements that it contains the nature and number of interrelations and the number of levels of embeddedness. The theory of complexity emphasizes that the relationships in the complex systems such as organizations are not linear, and have a structure revealing unexpected results and arising choices in which the events cannot be predicted.

The complex sub-systems that educational institutions have can be seen in Figure 3.

When Figure 3 is looked into thoroughly, one can easily see that the backgrounds of students may be totally different even if they appear to be similar. As can be seen in the above figure, each student may have a different family, social circle or school

Figure 3. The complex sub-systems in educational institutions

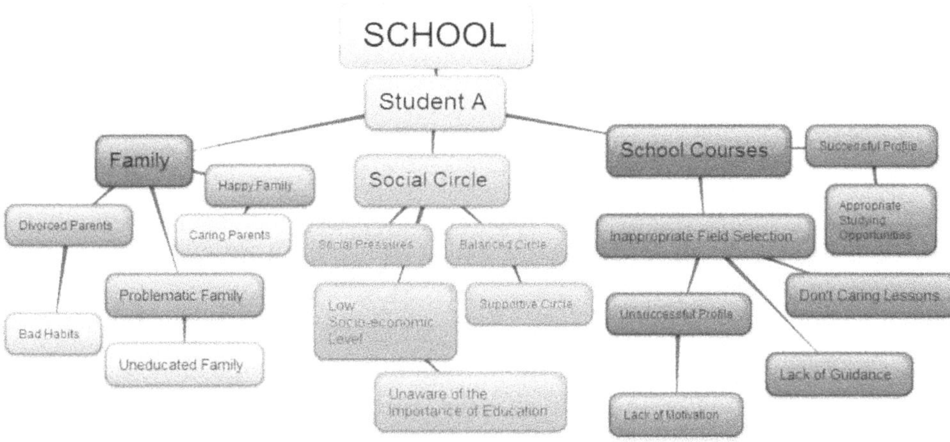

courses at different. The background of each student in the level of family, social circle and school courses divides into branches and become more complicated. As an example, the family background of a student may become more complicated in terms of living together or being divorced of the parents. Moreover, the background of each student in terms of family, social circle and school courses have a mixed connections among themselves. To give an example, the stability of a student's family is influential dramatically on the school course success. Similarly, the incoherence between the social circle and family of the student may affect all the subsystems in the background of the student. These examples show that the each subsystem in the background of a student has a contact with the other sub-systems; each system has a contact with the other systems and all the systems that form the background of a student have numerous close relationships between themselves. The balance of the relationships between the background units of a student is extremely crucial and it can change in a short time. The example given above in Figure 3 is directly associated with the principles of chaos and complexity theory. These principles are that it "operates in a non-linear way, that it is iterative (the output of one cycle becomes the input of the next) and that small variations in initial conditions lead to large differences in outcomes (Cunningham, 2000)". In his sense, the interrelated and branched structure of a student's background units can be explained with the first principle above. Furthermore, in educational institutions there are iterations. The reason for a student's failure at school may be directly associated with the family occasions of that student. This can be explained by the iterations, the second principle. Last but not least, the educational institutions and sub-systems in them have the characteristics of sensitivity to initial conditions. Each sub-system comprising family, social circle and school courses given in Figure 3 as background of a student has the clues of sensitivity to initial conditions. When Figure 3 is looked into, there is a sub-unit named "inappropriate field selection". The guidance of the family, teacher or manager with empathy in the stage of selecting the field to be studied can affect the whole chain, from farthest point to the nearest and can reverse all of the possibilities. The complicated characteristics of educational institutions require a genuinely effective manager or leader. Concerning this issue, Ellis (2005, p. 13) states that "managers must be able to deal with the complexity and speed of change that is occurring in the organization". Similarly, Tuncer and Açıkalın (2013, p. 2) also utters that "powerful change has always brought chaos and challenges". As a result, the managers who have the ability of coping with chaos, complexity and speed of change are required to have the ability of identifying the emotions truly because the emotions may have some clues about the chaos, complexity and speed of change.

FUTURE TRENDS

In the present age, in which the world is characterised as a small village, people do not perform only linear activities. Routinized life as a result of performing only linear activities is boring people. For this reason, now people are preferring to live a more complex way instead of living from home to work from work to home. This change is affecting organizations in such a way that the new life style, which eliminates the monotony, is showing itself in organizations. The quickly dissemination of the information in everyday life is making it rapid and ample in organizations, too. The inevitability of passing from hierarchical organizations to complex ones, which enables cross-functional teams to work in a crosswise manner in a wide area, is showing itself. The importance of interdisciplinary works is increasing. All these tendencies are seeming to continue rapidly in the future. The change in the social life and organizations is pushing the managers to have distinctive competencies different from the previous year. The emotional intelligence behaviours of a leader, who has the ability to adapt the change and use it on behalf of the goals, are coming forth as an effective quality about the future.

CONCLUSION

In this chapter educational institutions, leadership and emotional intelligence concepts in the context of chaos and complexity theories are scrutinised theoretically and practically. The unique features of educational institutions, the role of the leaders and the reflection of their emotional intelligence levels to the educational organizations are discussed deeply in the context of chaos and complexity theories. From the discussions that have been carried out so far, the authors find it possible to conclude that leaders' emotional intelligence levels in chaotic and complex situations have foremost effect on leadership in educational institutions. Besides, the authors of this chapter think that emotional intelligence levels can have a role in the rising of new leaders. On the basis of the finding discussed in this paper, it can be advised that certain activities that will enhance the leaders' level of emotional intelligence may be useful to educational institutions.

REFERENCES

Açıkalın, Ş. N., & Bölücek, C. A. (2014). Understanding of Arab Spring with chaos theory-uprising or revolution. In S. Banerjee (Ed.), Chaos theory in politics (pp. 29-47). Springer.

Allen, P. M. (2001). A complex systems approach to learning in adaptive networks. *International Journal of Innovation Management*, *5*(02), 149–180. doi:10.1142/S136391960100035X

Altunoğlu, S. (2009). Kaosun Keşfi ve Felsefesi, Bibliotech. *Felsefe-Sosyal Bilimler Dergisi, Eylül/Ekim* (9).

Ashforth, B. E., & Humphrey, R. H. (1995). Emotion in the workplace: A reappraisal. *Human Relations*, *48*(2), 97–125. doi:10.1177/001872679504800201

Bar-On, R. (2006). The Bar-On model of emotional-social intelligence (ESI). *Psicothema*, *18*(1), 13–25. PMID:17295953

Bass, B. M., & Bass, R. (2009). *The Bass handbook of leadership: Theory, research, and managerial applications*. Simon and Schuster.

Beyer, J., & Niño, D. (2001). Culture as a source, expression, and reinforcer of emotions in organizations. In R. L. Payne & C. L. Cooper (Eds.), *Emotions at work: Theory, research, and applications for management* (pp. 173–197). Chichester, UK: Wiley.

Burns, T. E., & Stalker, G. M. (1961). *The management of innovation*. London: Tavistock.

Bursalıoğlu, Z. (2013). *Okul yönetiminde yeni yapı ve davranış*. Ankara: Pegem Akademi.

Cartwright, D., & Zander, A. (1958). *Group Dynamics*. Evanston, IL: Row-Peterson.

Clapham, C., & Nicholson, J. (2009). *The concise Oxford dictionary of mathematics*. Oxford University Press. doi:10.1093/acref/9780199235940.001.0001

Cooper, R. K. & Sawaf, A. (1998). Executive EQ: emotional intelligence in leadership and organizations. New York: Berkley Publishing Group.

Cunningham, R. (2000). *Chaos, complexity and the study of education communities*. Paper presented to the British Educational Research Association Annual Conference, University of Leeds.

Daft, R. (2014). *The leadership experience*. Cengage Learning.

Duncan, D. (2002). Emotionally intelligent female primary head teachers: The new women of power? *Education*, *30*(3), 48–54.

Ellis, C. W. (2005). *Management skills for new managers*. New York: AMACOM.

Erçetin, Ş. Ş., Açikalin, S.N., & Bülbül, M.S. (2012). A multi-dimensional approach to leadership in chaotic environments. In S. Banerjee (Ed.), Chaos and Complexity Theory for Management: Nonlinear Dynamics: Nonlinear Dynamics (pp. 89-104). IGI Global.

Erçetin, Ş. Ş., Bisaso, S.M., & Saeed, F. (2015a). Understanding chaos and complexity in education systems through conceptualization of fractal properties. In Ş. Ş. Erçetin &S. Banerjee (Eds.), Chaos, Complexity and Leadership 2013 (pp. 147-161). Springer International Publishing.

Erçetin, Ş. Ş. (2015c). Understanding Recep Tayyip Erdoğan and Turkey with Plasma as a metaphor of the fourth state of matter. In Ş. Ş. Erçetin (Ed.), *Complexity and Leadership 2014* (pp. 1–13). Springer International Publishing.

Erçetin, Ş. Ş., & Bisaso, S. M. (2015b). The incorporations of fractals into educational management and its implications for school management models. In Ş. Ş. Erçetin & S. Banerjee (Eds.), *Chaos, Complexity and Leadership 2013* (pp. 35–55). Springer International Publishing. doi:10.1007/978-3-319-09710-7_4

Erçetin, Ş. Ş., & Düzer, A. (2008). Multidimensional perceptual leadership model. *World Applied Sciences Journal*, *3*(1), 25–33.

Erçetin, Ş. Ş., Potas, N., Kısa, N., & Açıkalın, Ş.N. (2013). To be on the edge of chaos with organizational intelligence and health. In S Banerjee (Ed.), Chaos and complexity theory for management: Nonlinear dynamics (pp. 184–203). Hershey, PA: IGI Global.

Ferreira, M. A. M., Filipe, J.A.C.B., Coelho, M., & Pedro, I.C. (2014). Chaos in World Politics: A Reflection. In S. Banerjee (Ed.), Chaos Theory in Politics (pp. 89-104). Springer.

Follet, M. P. (1940). *Papers on dynamic administration*. New York: Harper.

Goleman, D. (2013). İş başında duygusal zekâ (D. Balkara, H. Deniztekin, & O. Deniztekin, Trans.). İstanbul: Varlık Yayınları. (Original work published 1998).

Kayman, E. A. (2014). Chaos in education as an intelligent complex adaptive system. In Ş. Ş. Erçetin & S. Banerjee (Eds.), *Chaos and Complexity Theory in World Politics* (pp. 299–314). Global. doi:10.4018/978-1-4666-6070-0.ch019

Koçak, K. (2000). Kaos ve Atmosfer. *TÜBİTAK Bilim ve Teknik*, *391*, 94–97.

Lorenz, E. N. (1995). *The essence of chaos*. University of Washington Press.

Lucia, A. (1997). Leaders know how to listen. *HRFocus*, *74*(4), 25.

Maddock, R. C., & Fulton, R. L. (1998). *Motivation, emotions, and leadership: The silent side of management*. Greenwood Publishing Group.

Mandelbrot, B. B. (1983). *The fractal geometry of nature*. W. H. Freeman and Company.

Marion, R. (2006). Complexity in organizations: a paradigm shift. In A. Sengupta (Ed.), *Chaos, Nonlinearity, Complexity* (pp. 247–269). Springer Berlin Heidelberg. doi:10.1007/3-540-31757-0_9

Mason, M. (2008). Complexity theory and the philosophy of education. *Educational Philosophy and Theory*, *40*(1), 4–18. doi:10.1111/j.1469-5812.2007.00412.x

Mayer, J. D., & Salovey, P. (1990). Emotional intelligent. Bywood Publishing Co., Inc.

McMillan, E. (2004). *Complexity, Organizations and Change*. London: Routledge, Taylor and Francis Group.

Morgan, G. (1998). *Images of organization*. SAGE Publications.

Özdemir, S., & Kılınç, A. Ç. (2014). Değişen toplum ve okul. In. S. Özdemir (Ed.), Türk eğitim sistemi ve okul yönetimi (pp. 75-91). Ankara: Pegem Akademi.

Schlechty, P. C. (2011). *Okulu yeniden kurmak* (Y. Özden, Trans.). Ankara: Nobel Yayıncılık.

Sengupta, N. (2009). *Emotional intelligence*. New Delhi: Excel Books.

Sieben, B., & Wettergren, A. (Eds.). (2010). *Emotionalizing organizations and organizing emotions*. Palgrave Macmillan. doi:10.1057/9780230289895

Şişman, M. (1994). *Örgüt Kültürü: Eskişehir il merkezindeki ilkokullarda bir araştırma*. Anadolu Üniversitesi Eğitim Fakültesi Yayınları.

Snyder, K. J., Acker-Hocevar, M., & Wolf, K. M. (1995). *Chaos Theory as a Lens for Advancing Quality Schooling*. Academic Press.

Stogdill, R. M. (1948). Personal factors associated with leadership: A survey of the literature. *The Journal of Psychology*, *25*(1), 35–71. doi:10.1080/00223980.1948.9917362 PMID:18901913

Tetenbaum, T. J. (1998). Shifting paradigms: From Newton to chaos. *Organizational Dynamics*, *26*(4), 21–32. doi:10.1016/S0090-2616(98)90003-1

Titcomb, T. J. (1998). *Chaos and complexity theory (No. 9807)*. American Society for Training and Development.

Tuncer, A. M., & Açıkalın, Ş. N. (2015). Solution to chaotic situations in higher education: New generation universities as intelligent organizations. In Ş.Ş. Erçetin & S. Banerjee (Eds.), Chaos, Complexity and Leadership 2013 (pp. 1-9). Springer International Publishing.

Uğurlu, Z. (2015). Okul örgütü ve yönetimi. In *Türk eğitim sistemi ve okul yönetimi* (pp. 181–237). Ankara: Pegem Akademi.

Ural, S. S., & Açıkalın, Ş., N. (2015). Protection of fundamental rights and freedoms as a universal problem and chaos theory. In Ş.Ş. Erçetin & S. Banerjee (Eds.), Chaos, Complexity and Leadership 2013 (pp. 197-202). Springer International Publishing.

Wheatley, M. J. (2010). *Leadership and the new science: Discovering order in a chaotic world*. San Francisco: Berrett-Koehler Publishers.

Zerbe, W. J., Härtel, C. E., & Ashkanasy, N. M. (Eds.). (2010). *Emotions and organizational dynamism* (Vol. 6). Emerald Group Publishing. doi:10.1108/S1746-9791(2010)6

ADDITIONAL READING

Banerjee, S., & Erçetin, Ş. Ş. (2014) Chaos, Complexity and Leadership 2012. New York: Springer Proceedings in Complexity http://www.springer.com/us/book/9789400773615

Banerjee, S., Erçetin, Ş. Ş., & Tekin, A. (2014) Chaos Theory in World Politics (Vol-1). New York: Springer Understanding Complex Systems Series http://www.springer.com/us/book/9789401786904

Beyazyüz, M., & Erçetin, Ş. Ş. & Albayrak, Y. (2014). Crisis, Crisophilia and Crisophilic. In Ş.Ş. Erçetin and S. Banerjee (Eds.) Chaos and Complexity Theory in World Politics IGI Global, USA.

Erçetin, Ş. Ş. (Editor) (2012). Discussions about Leadership in Different Fields. USA: iUniverse Inc.

Erçetin, Ş. Ş. & Banerjee, S. (2014). Chaos and Complexity Theory in World Politics. IGI Global, USA. http://www.igi-global.com/book/chaos-complexity-theory-worldpolitics/99516

Erçetin, Ş. Ş. & Banerjee, S. (2015). *Chaos, Complexity and Leadership 2013*. Springer Proceedings in Complexity. New York: USA http://www.springer.com/us/book/9783319097091

Erçetin, Ş. Ş. (2016). *Chaos, Complexity and Leadership 2014*. Springer Proceedings in Complexity. New York: USA http://www.springer.com/us/book/9783319186924

Fineman, S. (Ed.). (2000). *Emotion in organizations*. Sage.

Gleick, J. (2014). Kaos: Kaosun kutsal kitabı (İ., A. Demir, Trans.). İstanbul: Alfa

Rickles, D., Hawe, P., & Shiell, A. (2007). A simple guide to chaos and complexity. *Journal of Epidemiology and Community Health*, *61*(11), 933–937. doi:10.1136/jech.2006.054254 PMID:17933949

KEY TERMS AND DEFINITIONS

Chaos: Chaos is a state that has non-linearity inside. There is no stability in chaos and the situations in chaos cannot be predicted or predetermined.

Complexity: Complexity is a concept that is strongly depends on the initial conditions. Complexity is about the complex relationships between units and sub-units in an organization.

Emotional Intelligence: Emotional Intelligence (EQ) is a concept that takes one's own emotional awareness to the starting point and by means of this awareness one can understand the other people in the circle. Moreover, emotional intelligence can help people making use of the emotions while achieving their goals.

Leadership: Leadership isn't just affecting others. Leadership is about initiating the relationships that push the organizations to the aimed level with minor but effective interventions.

Chapter 5
The University System as a Mere Organized Conflict Environment

Yusuf Abubakar
Sokoto State University, Nigeria

Abubakar Mansur Danjeka
Federal College of Education, Nigeria

ABSTRACT

The university system is an organized anarchy. It is a system that brings different components from different dimensions with different focus sometimes conflicting one another in one place, in order to achieve a particular goal. This paper examines the university as a system that comprises different faculties, departments and units managing different disciplines and courses/subjects. Chaos and complexity nature of university was examined and results show that each faculty and department is taking a different direction without caring about others, and in the same faculty there are different departments handling different courses. And lecturers are using different methods in the conduct of their lessons and researches, and all these struggles are made towards achieving university objectives. The challenges faced by university system and possible solutions to overcome those challenges were discussed.

INTRODUCTION

A university is a Latin word "Universitas" meaning, "a whole" it is an institution of higher or tertiary education and research, which grants academic degrees in a variety of subjects and provides both undergraduate and postgraduate educations.

DOI: 10.4018/978-1-5225-0460-3.ch005

(Wikipedia, the free encyclopedia). Universities are pluralistic institutions with multiple, ambiguous and conflicting goals. They are professional institutions that are primarily run by the professionals (i.e. the academics) often in its own interests rather than those of the clients and they are collegial institutions which are mainly headed vice-chancellors/Rectors. University mandate to conduct original research requires resources, infrastructure, human capital, and a climate of inquiry protected by the principles of academic freedom. (Stelmach, 2012). The nature and character of the university as an academic organization entails the achievement and maintenance of a harmonious environment conducive for the working together of various groups of staffs and the management team for the attainment of preselected missions and objectives (Ndum & Stella, 2013).

The university system is an organized anarchy. It is a system that brought together different component with different functions in one unit to work together, as an entity. The structure of the university include decision making bodies, service departments and academic core. Under decision making bodies, there are different committees, different faculties and departments dealing with different disciplines and courses. For instance, faculty of education, faculty of law faculty of science and the faculty of engineering, under each faculty there are several departments handling different courses or disciplines. The focus of each faculty is different from that of others. Faculty of education is busy trying to ensure the production of professional teachers and educational administrators, while in the faculty of law is struggles to see that qualified and professional legal practitioners are produced, this is applicable to each of the faculties. Each department in the university there are different lecturers handling different courses and each has a different style of delivering his lecture and different style of conducting research, as there is no one best way of conducting lesson as well as research.

Likewise, the other complexity area of university system is students and lecturers do have different aims and objectives, for instance lecture is lecturing for pay while students' focus is on the knowledge they acquire and it claims that university is meant for students but they only have little or no say about the activities in the school. In the view of aforementioned series of chaos and complexity in the university system, this paper focuses on those chaos and complexity and possible recommendations that could be used to overcome possible challenges that may likely encountered by university management.

THE UNIVERSITY AS A SOCIAL SYSTEM

A social system is a bounded set of elements and activities that interact and constitute a Single social entity (Hoy & Miskel, 2008) a social system is a model of organization that possesses a distinctive total unity beyond its component parts, distinguished from its environment by a clearly defined boundary. It is subunits, elements, and subsystems that are at least interrelated within relatively stable patterns of social order. (Olsen, n.d) However, the University is a system social of interaction, it is an organized whole that comprised of interacting personalities bound together in an organic relationship. It could also be featured by an interdependent of parts, a clearly defined population, and different from the environment, a complex network of social relationships. Like any formal organization, University, call attention to both the planned and unplanned, the formal and informal aspects of organizational life. (Bush, 2003).

The following are some of the indicators that made a university to be a social system, they are derived from basic assumptions of the social system theory as stated by Hoy & Miskel (2008).

- The University is a goal oriented and it has a multiplicity of goals. Besides preparing students for adult roles, there may be other goals at faculties, departments and individual levels.
- The University comprised of interdependent parts, characteristics, and activities that contribute to and receive from the whole. When one part is faulty, all the system will be affected.
- The University is peopled. People act in the roles of administrators, lecturers, Students and custodians, i.e. people is the most in important component in the university.
- The University is formally structured. Different components are put together to carry out specific functions and allocate resources. To some extent university is structured in form of bureaucratic setting i.e. in a hierarchical order.
- The University in normative. It is normative in the sense that, each person within the university is expected to behave in a particular manner. Formal rules and regulations and informal guides prescribe appropriate behavior.
- The university is sanction-bearing. The norm for behavior are enforce with reward and punishment. Formal mechanisms include expulsion, suspension, termination and promotion systems. While informal methods include the use sarcasm.
- The university is also an open system. The external environment typically supplies input to the system; there is a symbiotic relationship between university and external environment. It is affected by the values of the community, politics and history and its products will go back to the society.

THE UNIVERSITY STRUCTURE

Manga (2010) stated the following as the structure of the ideal University; namely Decision making bodies, Service department and the Academic core.

- **Decision Making Body**: In the university is made up of committees, departmental boards, faculty boards, the senate (Academic Boards), council and congregation.
- **Committees**: These are a body of persons appointed for a special function. At university there are several committees that help in decision, such as timetable committee, examination committee, Disciplinary committee welfare committee among others.
- **Departmental Boards**: These are the basic teaching and research units of the University. It is perhaps the most important unit as it is the entry point for both staff and students into the institution and it also the focal point of intellectual activity.
- **Faculty:** The faculty is an academic and administrative unit of the university, it is a collection of department that have similar intellectual and academic pursuits. For instance, some universities have the faculty of education, faculty of law, faculty of science, etc.
- **Senate / Academic Board**: The senate is the supreme governing body of the university in all academic matters. The vice-chancellor/Rector is the chairman of senate as is the chief academic officer of the university. Membership of the senate is made up of professors holding established chairs, All Deans of the senate, All Deans of faculties and directors of institutions, Representative of the alumni association, the Registrar, Deputy Vice-Chancellors or Deputy Rectors as the case maybe and a representative of the congregation. The senate has exclusive and highest authority in all academic affairs.
- **Council:** This is the executive governing body of the universities. The council delves into matters of financial and general operations of the institution. It exercises general control on finances, discipline, business, works and other services rendered in the university. It also makes university regulations that guide the policies.
- **Congregation:** This is a larger assembly of senior members of staff of the university who are holders of Ph,D degrees, both academic and professional. One member of congregation represents others in the senate and council.
- **Service Departments**: These are the departments that facilitate teaching and learning by providing essential services to the university community. The following are the services departments; Registry, the Bursary, Works and health services Departments.

- **Registry:** The registry is headed by the Registrar, who is the scribe of the institution and who is assisted by the principal Assistant Registrar, senior Assistant Registrar, Academic officer, other officials and clerical staff in the various departments of the registry. Sections under registry are: the General and Administrative, the Academic Office, the Establishment section, Student Affairs Office, Under student affairs office are the following units and they include; career and Counselling Unit, Transport Unit, Catering Services Unit, Hall of Residence unit, Health Services Unit, and Student Welfare Unit among others.
- **Bursary:** The bursary department handles the financial transactions of the university. The bursary is headed by a bursar who is assisted by a Deputy Bursar, Internal Auditors and other finance staff. The bursary serves as guarantor to staff loans and approves loans give to staff of the university as well as ensures prompt payment through making appropriate deductions. The bursar is both a member of the senate and of the council.
- **Works:** The works department is headed by the Estate Officer assisted by a team of engineers in various aspects of construction projects and maintenance of the school plant.
- **Health Services:** The health services department is headed by a Medical Director who is assisted by a team of medical doctors and consultants from various specializations, nurses, midwives, health technologists, laboratory attendants, pharmacists, ward servants and clinical staff. Health services department is the in charge of health care management of both staff and students.
- **Academic Core:** The academic core is made up of departments, institutes, centres and faculties. They focus mainly on teaching, research and community services and they teach all the subjects offered in the university. Their services include conducting of examinations and graduates Students for certification, supervise students' research projects, engages in national and international publications, organizes and participates in national and international conferences, workshops, distance learning and maintains national and international fellowships with other universities through joint research work sabbatical visitation and student exchange programs.

CHAOS AND COMPLEXITY IN THE UNIVERSITY SYSTEM

Murphy (2011) states that, Chaos theory can be generally understood as the study of complex systems, in which nonlinear problems are considered in their own right, rather than as inconvenient deviations from linearity. He further states that, within chaos theory, two general emphases exist. In the first, chaos is seen as the order's precursor and partner, rather than as it opposite. The focus here emphasizes on the spontaneous emergence of Self organization from chaos. The second branch the hidden order that exists within chaotic systems. Chaos in this usage is distinct from true randomness, because it can be shown to contain deeply encoded structures called "strange attractors." Whereas truly random systems show no discernible pattern when they are mapped into phase space, chaotic systems contract to a confined region and trace complex patterns within it (Hayles, 1990)

Therefore, system is simply the name given to an object studied in some field and might be abstract or concrete; elementary or composite; linear or nonlinear; simple or complicated; complex or chaotic. *Complex* systems are highly composite ones, built up from very large numbers of mutually interacting subunits (that are often composites themselves) who's repeated interactions result in rich, collective behaviour that feeds back into the behaviour of the individual parts. *Chaotic* systems can have very few interacting subunits, but they interact in such a way as to produce very intricate dynamics. *Simple* systems have very few parts that behave according to very simple laws. (J. Epidemiol Community Health, 2007).

NATURE OF CHAOS IN EDUCATIONAL INSTITUTIONS

Chaotic systems take complex forms, making their precise measurement difficult if not impossible (Glickman, 2001). Different measurement instruments have been put in place to evaluate and compare the performance of a school. However, due to the complex nature of schools, none of these assessment methods seem to measure precisely the school performance and have very limited validity for the following reasons (Cunningham, 2000):

However, Ambiguity models stress, uncertainty and unpredictability in organizations. These theories assume that organizational objectives are problematic and that institutions experience difficulty in ordering their priorities. The subunits are portrayed as relatively autonomous groups, which are connected only loosely with one another and with the institution itself. Decision-making occurs within formal and informal settings where participation is fluid. Ambiguity is a prevalent feature of complex organizations such as schools and is likely to be particularly acute during periods of rapid change (Bush, 2003)

Ambiguity models assume that turbulence and unpredictability are dominant features of organizations. There is no clarity over the objectives of institutions and their processes are not properly understood. Participation in policy making is fluid as members opt in or out of decision opportunities.

The Nature of University as a complex organization, highlighted by Manga (2010):

- The university has undefined, inconsistent and unclear goals. Where teaching, research and community service are not well conceptualized. The lecturer can teach the way he/she likes, do his/her research as he/she wants and serve the community, educational institution in his/her own matter of interest.
- The technology is undefined. Different methods are used in research. Some institutions use animals, some use human beings, and others use non-living objects. There is no one defined way of doing things.
- There is fluid participation. For instance, students graduate every year and replaced by new ones. There is no stability in this process and also student body is the largest body in the university, but they have little or no say in the decisions of the university as they may not be consulted in the construction of their facilities.
- In the university Decisions are made by the different committees and nobody has absolute power among HOD, Dean and registrar. Every correspondence is channeled to the registrar, but he himself has no vote. Thus everything is loosely coupled.

Furthermore, Kara (2008) states the following;

- The prior achievement of pupils is not taken into account and this is a major factor in pupil achievement at a later stage.
- Schools are differentially effective in different subjects and with pupils of different ability, which is not reflected in a single figure.
- Schools change over time; however, the achievement data used reflects only one group and is essentially historical data.
- Student mobility between schools is not reflected in the assessment.
- Social factors, sex of students, ethnic origin and social background are not taken into account. These factors are out of the school's control.

Therefore, assessing school performance and comparing one to another have become increasingly difficult given the complex nature of today's schools.

THE CONFLICT IN UNIVERSITY

Conflict is inevitable among humans. When two or more social entities (i.e., individuals, groups, organizations, and nations) come in contact with one another in attaining their objectives, their relationships may become incompatible or inconsistent. The conflict theory believes that there are always conflicts and competition in the university (Manga 2010). Changes take place mostly through conflict. In the university there are power configurations such as the students Union, Academic Staff Union and Non-Academic Staff Union among others. All of them, pursue to influence the university decisions in their own favour or interest.

The theory also postulated that, in formal organization like a University opportunity for conflict exists among the major dimensions of the system. This involved conflicts between role and norm, role and personality, role and goal, norm and personality, norm and goal, as well as goal and personality (Hoy & Miskel, 2008). There is also conflict between the system component and parts of the environment. For instance, conflict can arise between bureaucratic role and expectations and the value structure of the community.

Basically the following are the major types of conflict in the university:

- **Personal Conflict:** This arises from basic incompatibilities in the need structure of individuals. For example, an administrator or lecturer in a university may have strong needs for both security and may find it difficult to display appropriate organizational behavior of the university.
- **Goal Conflict:** This is common in organizations that have multiple goals that are often contradicting like a university where different faculties and departments struggle to attain certain goals.
- **Role-Goal Conflict:** This is where the bureaucratic expectations are inconsistent with organizational goals. Example, a lecturer may exhibit certain behavior like truancy, immorality, etc. which may result a problem towards achieving university goals and/or enforcing a particular dress code may hamper the stated goal of developing the whole child.
- **Role-Personality Conflict:** This occurs because individuals sometimes find themselves in roles for which they are not personality suited like assigning a role of Dean or HOD to a person who too considerate and lenient.
- **Goal-Personality Conflict:** This conflict highlights the difficulty the formal organization may have in getting individuals to embrace organizational direction. Teachers may for example not comply with organizational goals for a variety of individual reasons.
- **Norm-Goal Conflict:** This conflict may occur between the informal organization and goals. Informal norms may undermine the official goals of the organization. When lecturers are not sympathetic to the direction of the university, they resist, ignore, and sometimes sabotage it.
- **Role-Norm Conflict:** This occurs when there is tension between the formal and informal organization. For example interest group like academic staff union and university administrators.

Other issues that bring about conflict are highlighted as follows;

- **Change in Delegation:** Change in delegation can cause conflict. For example, when the authority of a manager is reduced, there can be intrapersonal conflict. The manager can perceive himself or herself to be less important to the organization, and this perception can cause severe emotional disturbance.
- **Change in Status:** Changes in status, especially when they disturb a given accepted hierarchy, can bring conflict in virtually every form possible. For example, such simple things as changes in title can be a significant cause of conflict. If a teacher perceives that the change in title from head of department to class teacher means a reduction in status, he or she is bound to feel disappointment and perhaps even resentment.
- **Change in Goals:** A change of goals or mission can also bring conflict to an organization's members. For instance, conflict would probably occur between members of a university faculty if the primary goal of the university were changed from a research to a teaching orientation. Research-oriented faculty would be in a conflict with those who place teaching as their top priority.
- **Organization Overlap:** When two or more people are assigned to carry out the same task, organization overlap occurs. Overlap causes interpersonal strain or conflict almost by definition. If teachers were both assigned the same task, there would undoubtedly be situations in which conflict could occur.
- **Resource Competition:** When the organization's resources are insufficient to meet the requirements of the subunits to do their work, there is *competition for scarce resources* (for example, budget allocations, assigned teaching positions, space of facilities). Two or more people or groups who are competing for the same resources or rewards are thrown into conflict as a result.
- When one party seeks to control the activities "belonging" to another unit (and the second unit seeks to fend off such "interference"), the issue is *autonomy* (for example, protecting one's "turf").
- When two parties in the organization must work together but cannot agree on how to do so, the source of conflict is *goal divergence* (for example, the school principal and the director of special education have differing views as to how mainstreaming issues are to be settled).

CHALLENGE OF UNIVERSITY SYSTEM

The following are some of the challenges that could be found in the university system as outlined by Stoner (n. d).

- **Misaligned Goals:** In a complex organization like university there is a misaligned goals which seems to be as one of the most difficult challenges occurs when people working on the same project have different goals. Working at cross-purposes, they often end up in interpersonal conflict. Conflicting goals stem from lack of a shared vision for the larger picture and are a setup for power struggles.
- **Conflicting Loyalties:** Another challenge in the university system is conflicting loyalties. When people are oriented along reporting lines rather than around the work, they don't see themselves as part of the same team. When the focus is on power and control instead of the goal, attention is turned in the wrong direction. Conflict between line managers and functional managers erupts around prioritizing projects and allocating resources.
- **Confusion about Roles and Responsibilities:** Due to the university system sometimes there is confusion about roles People don't know who to contact for information and/or fail to share important information with those who need it. Someone might be seen as dropping the ball when they never knew the ball was in their court. Frustration arises when projects get stalled or are not implemented as a result of lack of communication and coordinated efforts.
- **Delayed Decisions:** In a complex environment like university where several teams are working on different aspects of a project, it can be unclear who has the final authority over decisions, even among the decision makers. And it can be especially challenging in organizations where leaders are unaccustomed to sharing decision-making. Often everyone owns a piece of a project, but it's not clear who is accountable for the whole and as a result no one is steering the ship.
- **Managing a Diverse Workforce:** In a university a manager is dealing with diverse workforce which include different individuals and groups with differences in employment needs, work styles and aspirations.
- **Quality Assurance:** One of the greatest challenges facing universities is quality assurance. With globally competitive markets, a lack of quality assurance translates to poor products, inefficiency, loss of client and consequently poor business. The factors that are likely responsible for the challenge in quality assurance, rapid increase in teacher-student ratio in the university, decrease in contact between lecturer and students due to the technological advancement, demand is greater than supply of quality lecturers, increasing in motivation and commitment to excellence among academic staff and also universities seek to attract more students for profit maximization or other reasons. (Johnson, Hourizi & Forbes, 2009).

- **Academic Mobility:** Academic mobility has been in existence throughout the history of higher education and has important particularly for research and knowledge exchange. However, it recently come to be associated with commoditization and sometimes commercialization of higher education with technological advancement in information and transport, those who produce, package and disseminate knowledge as well as those who consume it (students and academic staff) increasingly move across the global knowledge market. Basically, student mobility is guided by consideration of quality, cost and length of programs abroad. And for academic staff, mobility is exercised in a search for better environment and in of service amongst other reasons. (Johnson, 2009),
- **Commercialization and Commoditization of Higher Education:** With the recognition of information as legal tender, universities are no longer mere repositories, disseminators and creators of knowledge but much more commercial institutions, that produce and transact knowledge as a commodity. In the current knowledge-based economy, universities are increasingly becoming running on business modes so that university managers are increasingly becoming powerful managers of human resource that produce, store and disseminate knowledge hence, for academic staff working in various universities, it is no longer "publish or perish" but " innovate or perish". Thus number of students is increasingly high. (Johnson, 2009).

CONCLUSION

The chapter examined the nature of university system as an organization that encompasses many component with different objectives in one place to work together. For instance, the university structure involves different faculties and departments, several committees tackling different issues concerning students and staff. Lecturers and students in university have different aims and objectives. Different methods of delivery and conduct of research, in the faculties of Arts and Humanities are dealing with social science related issues while faculty of sciences are focusing towards natural science issue, all in the same university.

KEY TRENDS IN OVERCOMING THE CHALLENGES

As a manager in a university should do the following in order to overcome the aforementioned challenges that are likely to be found in the university system. Alabi, (nd), postulated that, Clarification of Goals and Objectives: The clarification and continual refinement of goals and objectives, role definitions and performance standards help to avoid misunderstanding and conflicts. Focusing attention on super-ordinate goals that are shared by parties in conflicts may also help to defuse hostility and lead to more cooperative behaviour. Providing Valid Information and Avenue for Expression of Views. Information is needed to avoid blocking of communication flow that may lead to differences in perceptions. Effective management information system (MIS) is essential to provide requisite information which minimizes delays and ensures maximum utilization of resources. Most African universities lack effective computerized management information system for capturing, processing, storing, retrieving and disseminating relevant information (Alabi, 2000).

In addition to that, a university manager should;

- Make sure your goals are linked to the same vision and goals as others you are working with. If you are experiencing conflict, check to see if you share the same goals before you start blaming each other. Understand you are part of a team, whether or not you all report to the same person. If the team is not successful, you cannot be successful. Think "we" instead of "me."
- Focus on the work to be done and the best way to accomplish it. For each major project or key responsibility area, create a Team Charter that clarifies priorities, processes, decision making, resources, etc. Make sure your meetings are purposeful. Use a good project tracking system to avoid wasting time with meetings that only provide updates. Set an example and act in ways that encourage collaboration across boundaries.
- See beyond your own piece of work in the context of the bigger picture. Set up a communication plan. For each aspect of the project, identify what role each person plays: responsible to do the work and involved in decisions (several people). Ultimately accountable for success (usually just one person, although it can be shared). Consult to the project because they have needed information or expertise but do not make decisions or do the work. Informed, people who are not directly involved but need to be updated on progress.
- Find out everyone who needs to be involved in decisions. Make sure they all know and agree who is involved. Create a decision-making model that identifies who will make which kinds of decisions (similar to the RACI chart – in fact this will be easier to do if you have created a RACI chart). Keep decision-makers up-to-date so they have the information they need to make informed decisions. Give them advance notice that you will be bringing a decision forward. If you are put in a situation where you are accountable for the success of a project without the authority to make decisions on how to execute, push back or proceed carefully.

- A university manager needs to understand works as individuals and as a group, why because, they come from different background and they have different needs and aspirations and university has a desired goals to achieve, therefore every worker should be directed towards achieving those goals after satisfying their needs.
- In order for university to conform and meet up with global mobility, it should flexible in such a way that it can be adapted to the changing educational situation in world. And provides quality teachers who will be able to produce quality graduate as well as initiation of efficiency and effectiveness in the school programs.
- In order to support this situation in global market, there should be diversification of academic programs, in terms of structure, methodology and content. Likewise, in addition to traditional programs which involve immigration outside one's country for the purpose of study other programs like sandwich programs, study abroad programs and distance learning should be emerged.
- University should have a closely working relationship with other universities and local communities and industries and also review the university curriculum to march the labour market.

REFERENCES

Afzahur, R. M. (2001). *Managing conflict in organizations* (3rd ed.). Greenwood.

Alibi, A. T. (n.d.). *Conflict in Nigeria University: cause and management*. Academic Press.

Bush, T. (2003). *Theories of educational management* (3rd ed.). London: Sage Publishers.

European University Association. (2007). *Managing the University Community: Exploring Good practice*. Retrieved on 25th mar, 2015 through www.eua.be

European University Association. (2007). *Managing the university community: exploring good practice*. Retrieved from www. Eua.be

Ghaffar, A. (n.d.). Conflict in schools: Its causes & management strategies. *Journal of Management Sciences, 11*(1), 213-229.

Hoy, W. K., & Miskel, C. G. (2008). Educational administration: Theory, research and practice (8th ed.). New York. McGraw-Hill companies, Inc.

Institute for Higher Education Policy. (2001). College and university Ranking system. *Global perspectives and American challenges.* Retrieved from www.ihep.org

Kara, Y. (2008). *Understanding today's schools with chaos theory. Fountain on life, knowledge and belief.* Retrieved from http://www.fountainmagazine.com/Issue/detail/Understanding-Todays-Schools-with-Chaos-Theory

Manga, S. D. (2010). *Administration of higher education. Administration and planning.* Unpublished manuscript, Usmanu Danfodiyo University, Sokoto Nigeria.

Peter, A., et al. (2011). *Challenges for faculty managements at African higher education institutions.* Germany University of Applied Science, Osnabruck.

Poluku, K. (2013). Analysis of conflict management and leadership for organizational change. *International journal of Research in social science.* Retrieved from www.ijsk.org/ijras

Stoner, J. L. (2012). *Manage the challenges of working in a matrix organization.* Retrieved on 25th mar, 2015 from seapointcenter.com/matrix-leadership/

U S National Library of Medicine National Institute of Health. (2007). A simple guide to chaos and complex. *Journey of Epidemiology and Community Health, 61*(11), 933-937.

Chapter 6
Principals' Views on Stigma Tendency

Halime Güngör
Hacettepe University, Turkey

ABSTRACT

Stigma takes place as a concept in the literature that affects individuals' lives negatively as well as can disrupt chemistry of organization and inhibit organizational behavior. Stigma is a concept that ostracizing individuals or groups because of their features that are not accepted by society and restricted their life on all areas. Studies on stigma have been increased during twenty years. Although school principals play the lock role and the most affecting actor in positively or negatively to school, studies on principals' stigma tendency are extremely limited. Stigma tendencies of principals' can affect chemistry of organization much more than stigma tendencies of teachers. The purpose of the study is to determine stigma tendencies of principals' by using quantitative research methods. According to research results; principals' stigma tendencies are significantly different according to seniority, level of religious feelings and health status.

INTRODUCTION

Stigma takes place as a concept in the literature that affects individuals' lives negatively as well as can disrupt chemistry of organization and inhibit organizational behavior. Stigma tendencies of principals' can affect chemistry of organization much more than stigma tendencies of teachers.

DOI: 10.4018/978-1-5225-0460-3.ch006

According to Link and Phelan (2001) stigma is a force that allows the emergence of labeling, stereotypes, divisions, status loss and discrimination. Jones et al. (1984) defines stigma as relationship between an "attribute and a stereotype" to produce a definition of stigma as a "mark" (attribute) that links a person to undesirable characteristics (stereotypes). Elliott, Ziegler, Altman and Scott (1982) emphasized the social interaction in stigma and defines stigma as a form of deviance that leads others to judge an individual as illegitimate for participation in a social interaction. This is thought to affect a judgment on person's dangerousness and unpredictability. When one time individual are considered as illegitimate, it is not expected to show normal social behavior, that individual may be neglected or isolated from the group. An attribute of stigmatized person makes them different in the eyes of others and discredit the person. Furthermore, according to Parker and Aggleton (2003) stigma causes some groups to be devalued and others to feel that they are superior. To sum up, stigma is isolation of individuals or groups which have features that unacceptable in society and also restriction of individuals' or groups' lives in all areas.

When the history of stigma is explored, it is seen that stigma concept has been experiencing since the oldest societies. Ongoing caste system in India since ancient times can be an example of stigma. There have been stigmatizing attitudes of higher caste groups in interaction of scheduled and higher caste groups and stigma affect especially scheduled caste groups in terms of psychological, social and physical health (Jaspal, 2011).

However stigma to individuals that different from society is particularly more clear and obvious when diseases are in question. While epilepsy was known as feared disease in Ancient Greek and Rome (www.turkepilepsi.org.), leprosy and plague were also perceived as an evil given by God. People who suffered from syphilis which emerged in Europe are doomed and also patients with tuberculosis, with cancer, AIDS sufferers (Oran & Şenuzun, 2008) and the stigma against those with mental illness (Üçok, 2003) has a remarkable history. In addition, the stigma were first developed for people with mental illness in many societies (Kocabaşoğlu & Aliustaoğlu, 2003).

In 1930s there had been law regulations as restriction on not to got marriage, not to had children, given as adopt of epilepsy patients and also closure to mental health hospital to epilepsy patients who had even frequent bouts in the UK (www.turkepilepsi.org.). This situation can be considered as a proof that stigma of epilepsy. Likewise, hundred and eighty thousand psychiatric patients in Nazi Germany during six years between the years of 1939-1945 murdered under the supervision of psychiatrists, after Hitler came to power in 1933, he arranged the law for prevent of hereditary diseases, thus began the discriminatory process (Bilge & Çam, 2010).

In the 1980s, theory and research began to challenge traditional perspectives on how stigma affects self-esteem and academic performance, among other outcomes

(Major & O'Brien, 2005). Stigma with deep rooted history has been continuing to experience as damaging both individuals and society. When considering individuals from chaos theory; chaos can be defined as individual's getting out of the daily routine in spiritual, mental, physical, emotional or existing in another dimension of life, and also domination of uncertainty as well as instability to an individual (Güngör, 2014).

Although stigma is a phenomenon that has been experiencing all areas of life and all sectors in society and also there has been increasing research on stigma during last twenty years, studies on principals' stigma tendency is extremely limited. School principals are the most affecting actors more positively or negatively to educational organizations and also they play a key role in schools.

If a school principal behaves fairly, consistently, competent, be in the right place and at the right time, some issues can be solved before become problems and principals' this kind of behavior supported creation of a positive school culture (Balcı, 2001). Thus, according to Güngör's (2013) research results, when stigma tendencies of principals are decreased, school culture is strengthened; and also when teachers and principals views are compared, the finding show that principals' stigma tendency affect school culture more negatively than principals' stigma tendency. All in all, leaders are important actors in the context of effects of stigma to stigmatized person. Therefore, leaders held the power of increase or decrease the pressure on stigmatized individuals and also they are balance point in organizations. When considered in this respect, we cannot expect to determine in advance the moment when a society or organization will move into a chaotic state, which leadership process will arise, who will take on leadership or what the consequences will be (Erçetin et al., 2013a,b).

In this sense, it is clear that, principals' stigma tendency do harming effects on organizations and stigma tendencies of principals need to search. In this sense, this study's aim is to determine views of principals who have been working in elementary and high schools in Zonguldak, Turkey on stigma. In the study "What are the views of principals on stigma?", "Are views of principals on stigma significantly different according to gender, seniority, level of religious feelings, health status?" questions were searched.

METHODOLOGY

The purpose of the study is to determine stigma tendencies of principals' by using quantitative research methods. Relational survey model of causal-comparative research type was used in research. Survey model is aimed to describe a situation in the past or current by existing term (Karasar, 2006). Principals' views on stigma tendency were surveyed.

Research sample of the study is consisted of 125 administrators who have been working in public schools (elementary and high) in Zonguldak. Stratified sampling method from probability sampling methods was used to determine research sample. Stratified sampling is a sampling method that is guaranteed to be represented in the sample all subgroups of the universe in the research sample; sub-samples are combined to obtain a total sample by taking simple and unbiased sampling of each layer. Research universe consists of seven districts. Because of nature of the research universe, each district conceived as layer and intended to represent all layers in the sample. Sample size of districts are not equal, because analysis was not intended to be done in terms of districts, so research sample was selected by simple and random sampling method. Research sample is consisted of one hundred and twenty five principals. Characteristics of the research sample are given in Table 1.

According to Table 1 %83,2 of participants (n=104) are male and %16,82 of participants (n=21) are female. %4,8 of participants have 1–5 years of seniority (n=6), %7,2 of participants have 6–10 years of seniority (n=9), %24 of participants have 11–15 years of seniority (n=30), %16,8 of participants have 16–20 years of seniority (n=21), %47,2 of participants have 21 or more years of seniority (n=59). According to participants' own views, %4,8 of participants (n=6) have low level of

Table 1. Characteristics of the research sample

Characteristics of the Research Sample		**n**	**%**
Gender	Male	104	83,2
	Female	21	16,8
Seniority	1-5 years	6	4,8
	6-10 years	9	7,2
	11-15 years	30	24,0
	16-20 years	21	16,8
	21 years and above	59	47,2
Level of religious feelings	Low	6	4,8
	Medium	82	65,6
	High	20	16,0
	None of them	17	13,6
Health status	Some troubles with bodily health	14	11,2
	Some troubles with psychological health	2	1,6
	Have any health problem	109	87,2
Total		125	100

religious feelings, %65,6 of participants (n=82) have mid-level of religious feelings, %16 of participants (n=20) have high level of religious feelings and %13,6 of participants (n=17) have no religious feelings. %11,2 of participants (n=14) have been living some troubles with bodily health, %1,6 of participants (n=2) have been living some troubles with psychological health and %87,2 of participants (n=109) have any health problem.

Data Collection Tools

Stigma Scale which was developed by Yaman and Güngör (2013) and personal information form which was prepared by researchers were used for data collection. Five likert type scale was used (1 - strongly disagree, 2 - Disagree, 3 - Partially agree, 4 - agree, 5 - Completely Agree) in the scale. The lowest score that can be obtained from the scale is 22 the highest score is 110. According to Stigma Scale evaluation the stigma tendency of individuals who get below 55 point is low and the stigma tendency of individuals who get above 55 point is high (Yaman ve Güngör, 2013). Stigma scale is composed of 22 items and four subscales. Subscales are named as discrimination and exclusion, labeling, psychological health and prejudice. Factors are explained %13.4, %11.01, %10.02, %8.18 of the total variance respectively. Stigma Scale is explained %43.63 of the total variance. Four factors eigen values are 5.261, 1.80, 1.345, 1.194 respectively. There is no reverse rated items in scale. Cronbach alpha internal consistency coefficient of Stigma Scale was calculated as .84, Spearman-Brown correlation coefficient was calculated as .85 and Guttman split-half value was calculated as .85. Cronbach alpha internal consistency coefficient of Stigma Scale was calculated as .85 in this study.

Data Analysis

SPSS Package Program was used for statistical analysis of the data. Descriptive statistical analysis was used by analyzing of "What are the views of principals on stigma?" sub-problem. In the analysis of the "Are views of principals on stigma significantly different according to gender, seniority, level of religious feelings, health status?" sub-problem; independent sample t-test was used to determine the significance level in terms of gender, one-way ANOVA was used to determine the significance level in terms of seniority, level of religious feelings and health status. If there is significant difference as a result of variance analysis, Scheffe test was used in order to determine the source of the difference.

FINDINGS

Descriptive statistical analysis on principals' views about stigma is given in Table 2.

Table 2. Descriptive statistical analysis on principals' views about stigma

Item No	Subscales and Items of the Scale	N	$\overline{x}$	Ss	Min.	Max.
Discrimination and Exclusion		125	1.79	4.16	6.00	30.00
25	It is natural to discriminate employees	125	1.69	.919	1	5
26	People with different accents may be underestimated.	125	1.72	.848	1	5
27	People who wear contrary to society deserve exclusion	125	1.91	.933	1	5
28	Not seeing social acceptance of a person who has countryman style is natural	125	1.82	.925	1	5
29	Beautiful/handsome people are more successful	125	1.92	1.126	1	5
17	Someone who have different religious beliefs in the work environment disturbs me	125	1.72	.989	1	5
Labelling		125	2.33	4.26	6.00	26.00
7	If I had a choice I would not want to work with a person who was born as without marriage.	125	1.87	.950	1	5
8	I don't want to communicate with someone that I thought he/she has different sexual preferences	125	2.46	1.254	1	5
9	Older personnel's efficiency in the work environment is quite low	125	**2.75**	1.112	1	5
11	Female staff in work environment is more problematic	125	2.54	1.208	1	5
13	People can show different attitudes according to the origins of someone.	125	1.95	1.177	1	5
14	Divorced staff could have negative impacts to work environment	125	2.42	1.065	1	5
Psychological Health		125	**2.84**	3.44	5.00	21.00
18	Individuals who take psychological help have personality problems.	125	**2.61**	1.135	1	5
19	Trust decreases to people known to take psychological help.	125	**2.74**	.924	1	5
20	I would be disturbed in the same working environment with an individual who lives in contrary to society rules.	125	**2.98**	1.215	1	5
24	A person who has not social sharing with colleagues is problematic.	125	**2.93**	1.041	1	5
30	People who are in closed to communication is problematic.	125	**2.94**	1.145	1	5

continued on following page

Table 2. Continued

Item No	Subscales and Items of the Scale	N	$\overline{x}$	Ss	Min.	Max.
Prejudice		125	**2.72**	3.35	6.00	23.00
3	I don't want to work in the same environment with a person who is involved in criminal offense.	125	**2.98**	1.070	1	5
4	People who do not share the same world view behave unfair each other.	125	2.30	1.064	1	5
10	Employees with more seniority like to establish authority over the younger ones.	125	**2.69**	1.103	1	5
21	I don't meet outside work with a person whose lifestyle do not fit with me.	125	**2.86**	1.210	1	5
23	A person whose performance is begrudge by others is alone in the work environment.	125	**2.76**	1.058	1	5
Stigma Scale Total		125	2.42	11.66	26.00	82.00

According to Table 2, we can say that there are stigma tendency due to the principals views on "Older personnel's efficiency in the work environment is quite low ($\overline{x}$ =2.75)" item on labeling subscale, "Individuals who take psychological help have personality problems ($\overline{x}$ =2.61)", "Trust decreases to people known to take psychological help ($\overline{x}$ =2.74)", "I would be disturbed in the same working environment with an individual who lives in contrary to society rules ($\overline{x}$ =2.98)", "A person who has not social sharings with colleagues is problematic ($\overline{x}$ =2.93)", "People who are in closed to communication is problematic ($\overline{x}$ =2.94)" items in psychological health subscale ($\overline{x}$ =2.84) . There are stigma tendency due to the principals views on "I don't want to work in the same environment with a person who is involved in criminal offense ($\overline{x}$ =2.98)", "Employees with more seniority like to establish authority over the younger ones ($\overline{x}$ =2.69)", "I don't meet outside work with a person whose lifestyle do not fit with me ($\overline{x}$ =2.86)", "A person whose performance is begrudge by others is alone in the work environment ($\overline{x}$ =2.76)" items in prejudice subscale ($\overline{x}$ =2.72).

Analysis of principals' views on stigma in terms of gender is given in Table 3.

According to Table 3, views of principals in terms of gender is similar on *discrimination and exclusion, labeling, psychological health* and *prejudice* subscales, $t(123)=.945$, $t(123)=1.408$, $t(123)=1.698$, $t(123)=1.916$, $p>.05$. Analysis of principals' views on stigma according to seniority is given in Table 4.

According to Table 4, principals' views on stigma in terms of seniority are significantly different on discrimination and exclusion subscale $F(4,120)=3.896$, $p<.01$. According to Scheffe test results which is used to determine which years significantly different from each other in terms of seniority, stigma tendencies of principals who are 21 or more years of seniority ($\overline{x}$ =12.03) in discrimination and exclusion subscale are higher than principals who are 16-20 years of seniority ($\overline{x}$ =8.62).

Table 3. Results of t-test belongs to the perceptions of principals' views on stigma in terms of gender

Subscales of the S S	Gender	N	$\bar{x}$	S. S.	S. E. M.	F	t	Sd	p
Discrimination and Exclusion	Male	104	10.94	4.26	.4173	.005	.945	123	.346
	Female	21	10.00	3.67	.8017				
Labeling	Male	104	14.24	4.38	.4300	1.566	1.408	123	.162
	Female	21	12.81	3.44	.7515				
Psychological Health	Male	104	14.43	3.41	.3349	.042	1.698	123	.092
	Female	21	13.05	3.383	.7383				
Prejudice	Male	104	13.34	3.40	.3334	.546	-1.916	123	.058
	Female	21	14.86	2.85	.6221				
Total	Male	104	52.95	11.879	1.1649	.298	.801	123	.425
	Female	21	50.71	10.55	2.3033				

Views of principals in terms of seniority is similar on *labeling, psychological health* and *prejudice* subscales, $F(4,120)=.411$, $F(4,120)=.897$, $F(4,120)=.960$, $p>.05$.

Analysis of principals' views on stigma according to level of religious feelings is given in Table 5.

According to Table 5, principals' views on stigma in terms of level of religious feelings are significantly different on psychological health subscale $F(3,121)=4.044$, $p<.01$. According to Scheffe test results which is used to determine which levels significantly different from each other in terms of level of religious feelings, stigma tendency of principals who have mid-level religious feelings ($\bar{x}=14.77$) in psychological health subscale are higher than stigma tendencies of principals who have high level of religious feelings ($\bar{x}=12.25$) .

Views of principals in terms of level of religious feelings is similar on *discrimination and exclusion, labeling* and *prejudice* subscales, $F(3,121)=.043$, $F(3,121)=.790$, $F(3,121)=2.060$, $p>.05$.

Analysis of principals' views on stigma according to level of health status is given in Table 6.

According to Table 6, principals' views on stigma in terms of health status are significantly different on *general stigma tendency* $F(2,122)=3.561$, $p<.05$. According to Scheffe test results which is used to determine which status significantly different from each other in terms of health status, stigma tendencies of principals

Table 4. Results of One Way ANOVA belongs to the perceptions of principals' views on stigma in terms of seniority

Subscales of the S S	Seniority	N	$\bar{x}$	S. S.	Sd (Between G)	Sd (Within G)	F	p	Significant Difference
Discrimination and Exclusion	1-5 years	6	11.33	4.08	4	120	3.896	**.005**	21years and above-16-20years
	6-10 years	9	11.56	4.59					
	11-15 years	30	9.50	3.35					
	16-20 years	21	8.62	2.80					
	21 years and above	59	12.03	4.49					
	Total	125	10.78	4.16					
Labeling	1-5 years	6	12.67	3.33	4	120	.411	.800	_
	6-10 years	9	14.11	4.68					
	11-15 years	30	13.97	4.36					
	16-20 years	21	13.29	3.63					
	21 years and above	59	14.389	4.51					
	Total	125	14.00	4.26					
Psychological Health	1-5 years	6	12.33	4.03	4	120	.897	.468	_
	6-10 years	9	14.11	2.71					
	11-15 years	30	14.47	4.05					
	16-20 years	21	13.43	3.31					
	21 years and above	59	14.54	3.18					
	Total	125	14.20	3.44					
Prejudice	1-5 years	6	15.67	2.50	4	120	.960	.432	_
	6-10 years	9	13.89	3.69					
	11-15 years	30	13.20	3.21					
	16-20 years	21	12.91	2.74					
	21 years and above	59	13.78	3.62					
	Total	125	13.59	3.35					
Total	1-5 years	6	52.00	11.08	4	120	1.394	.240	_
	6-10 years	9	53.67	9.95					
	11-15 years	30	51.13	11.82					
	16-20 years	21	48.24	10.30					
	21 years and above	59	54.75	12.12					
	Total	125	52.58	11.66					

Table 5. Results of One Way ANOVA belongs to the perceptions of principals' views on stigma in terms of level of religious feelings

Subscales of the Stigma Scale	Level of Religious Feelings	N	$\bar{x}$	S. S.	Sd (Between Groups)	Sd (Within Groups)	F	p	Significant Difference
Discrimination and Exclusion	Low	6	10.33	3.93	3	121	.043	.988	–
	Middle	82	10.83	3.86					
	High	20	10.90	5.72					
	None of them	17	10.59	3.89					
	Total	125	10.78	4.16					
Labeling	Low	6	13.50	3.27	3	121	.790	.502	–
	Middle	82	14.40	4.46					
	High	20	13.50	3.93					
	None of them	17	12.82	3.99					
	Total	125	14.00	4.26					
Psychological Health	Low	6	12.00	2.00	3	121	4.044	**.009**	Middle-High
	Middle	82	14.77	3.37					
	High	20	12.25	3.31					
	None of them	17	14.53	3.37					
	Total	125	14.20	3.44					
Prejudice	Low	6	11.83	2.79	3	121	2.060	.109	–
	Middle	82	13.84	3.04					
	High	20	14.30	4.13					
	None of them	17	12.18	3.68					
	Total	125	13.59	3.35					
Total	Low	6	47.67	7.99	3	121	1.060	.369	–
	Middle	82	53.84	11.43					
	High	20	50.95	12.95					
	None of them	17	50.12	12.09					
	Total	125	52.58	11.66					

Principals' Views on Stigma Tendency

Table 6. Results of One Way ANOVA belongs to the perceptions of principals' views on stigma in terms of level of health status

Subscales of the Stigma Scale	Health Status	N	$\bar{x}$	S. S.	Sd (Between Groups)	Sd (Within Groups)	F	p	Significant Difference
Discrimination and Exclusion	Body Health	14	12.71	3.62	2	122	2.08	.129	_
	Psychological Health	2	13.00	9.89					
	Physical Disability	-	-	-					
	None of them	109	10.49	4.09					
	Body Health and Psychological Health	-	-	-					
	Total	125	10.78	4.16					
Labeling	Body Health	14	16.07	4.08	2	122	3.24	.042	
	Psychological Health	2	18.50	6.36					
	Physical Disability	-	-	-					
	None of them	109	13.65	4.17					
	Body Health and Psychological Health	-	-	-					
	Total	125	14.00	4.26					
Psychological Health	Body Health	14	15.50	2.53	2	122	1.22	.297	_
	Psychological Health	2	13.00	5.66					
	Physical Disability	-	-	-					
	None of them	109	14.05	3.49					
	Body Health and Psychological Health	-	-	-					
	Total	125	14.20	3.44					
Prejudice	Body Health	14	15.29	2.70	2	122	2.70	.071	_
	Psychological Health	2	16.00	1.41					
	Physical Disability	-	-	-					
	None of them	109	13.33	3.38					
	Body Health and Psychological Health	-	-	-					
	Total	125	13.59	3.35					
Total	Body Health	14	59.57	9.76	2	122	3.56	**.031**	Body Health-None of them
	Psychological Health	2	60.50	23.3					
	Physical Disability	-	-	-					
	None of them	109	51.53	11.4					
	Body Health and Psychological Health	-	-	-					
	Total	125	52.58	11.6					

who have some trouble with their bodily health ($\overline{x}$=59.57) are higher than stigma tendencies of principals who are healthy ($\overline{x}$=51.53).

Views of principals in terms of level of health status is similar on *discrimination and exclusion, labeling, psychological health* and *prejudice* subscales, $F(2,122)=2.084$, $F(2,122)=3.244$, $F(2,122)=1.225$, $F(2,122)=2.706$, $p>.05$.

DISCUSSION

Stigma tendencies of principals' were searched by using quantitative research methods with this study. According to research results, we can say that there are stigma tendency due to the principals views on "Older personnel's efficiency in the work environment is quite low" item on labeling subscale, "Individuals who take psychological help have personality problems", "Trust decreases to people known to take psychological help", "I would be disturbed in the same working environment with an individual who lives in contrary to society rules", "A person who has not social sharing with colleagues is problematic", "People who are in closed to communication is problematic", "I don't want to work in the same environment with a person who is involved in criminal offense", "Employees with more seniority like to establish authority over the younger ones", "I don't meet outside work with a person whose lifestyle do not fit with me", "A person whose performance is begrudge by others is alone in the work environment" items.

There was no significantly difference in gender according to principals' views on stigma tendency. Furthermore, according to Dalkıranoğlu and Çetinel's study's results (2008), principals' attitudes towards gender discrimination were found similar, but male managers' thoughts on female staff's working performance is that female staff's performance is more inadequate than male staff's performance.

According to research results, stigma tendencies of principals who are 21 or more years of seniority in discrimination and exclusion subscale are higher than principals who are 16-20 years of seniority in discrimination and exclusion subscale. In this case, it can be considered to experience the negative impact on organizational culture in organizations managed by the principals who are 21 or more years of seniority. Thus, according to a research result which was searched by Şimşek (2003), there was a high positive correlation between communication skills of school principals and school culture and this relationship was not differentiate according to gender of teachers but was differentiate according to seniority of teachers and teachers' working duration in their school.

In addition, stigma tendency of principals who have mid-level religious feelings in psychological health subscale are higher than stigma tendencies of principals who have high level of religious feelings. Thus, according to a research result which was searched by Varas-Díaz, Neilands, Malavé Rivera & Betancourt (2010), participants who identified that religion was important to them and participated in such activities showed the highest levels of HIV/AIDS stigma, on the contrary, participants who identified that religion was not important to them and not participated in such activities exhibited significantly less HIV/AIDS stigma. Furthermore, in doctoral dissertation of Brown (2004), relationship between religion and effects of religion on psychological health and spirituality in getting help for mental illness, the role of cultural mistrust and stigmatization were searched. According to analyzed results of data collection tools that were applied to 18-55 years old adults in Baptist and Methodist Churches where African-American people go, a high correlation was found between mental illness stigma and spirituality.

Stigma tendencies of principals who have some trouble with their bodily health are higher than stigma tendencies of principals who are healthy. People's stigmatization and judging to unlike him/herself is undoubtedly one of the biggest weaknesses of human (Zorlu & Çalım, 2012). Stigmatized feature is based on the social experience of the individual; stigma related to being disabled is not genetic or environmental but is caused by interaction. In terms of both environmental and cultural meaning there have been strong effects of emotional and behavioral reactions to disabled (Camara, 2011). Therefore, highness of stigma tendency of principals who have some trouble with their bodily health can be considered as caused by their psychological status.

SOLUTIONS AND RECOMMENDATIONS

Principals should be aware of their prejudices and their stigma tendency. It is very important that principals should develop conscious on harmful effects of principals' stigma tendency to organization. In this sense, awareness and experience groups can be helpful to decrease stigma tendency of principals. According to research results, although to manage people who have some trouble with their psychological health is an unwanted situation by principals, principals should help these personnel about to get psychological help for their psychological situation.

Highness of stigma tendency of principals who are 6-10 years and 21 or more years of seniority is one of the results of the research. Principals who are 21 or more years of seniority should be supported by seminars and educational activities to adapt the changes in today's conditions. Principals who are 6-10 years of seniority can be supported by experienced mentor principals in terms of sharing of experience. To the training of principal candidates in graduate educational management can take place in the regulations.

ACKNOWLEDGMENT

I want to thank with endless love to Prof. Dr. Şefika Şule Erçetin of Hacettepe University, Turkey for her affection, patience, support and everything that she gave us. She is the milestone in the life of me and my other friends.

REFERENCES

Balcı, A. (2001). *Etkili okul ve okul geliştirme*. Ankara: PegemA Yayıncılık.

Bilge, A., & Çam, O. (2010). Ruhsal hastalıklara yönelik damgalama ile mücadele. *TAF Preventive Medicine Bulletin*, *9*(1), 71–78.

Brown, J. A. (2004). *African American church goers' attitudes toward treatment seeking from mental health and religious sources: The role of spirituality, cultural mistrust and stigma toward mental illness*. (Unpublished doctoral dissertation). University of North Caroline, Chapel Hill, NC.

Camara, N. J. (2011). *Life after disability diagnosis: The impact of special education labeling in higher education*. (Unpublished doctoral dissertation). Georgia State University, Atlanta, GA.

Dalkıranoğlu, T., & Çetinel, F. G. (2008). Konaklama işletmelerinde kadın ve erkek yöneticilerin cinsiyet ayrımcılığına karşı tutumlarının karşılaştırılması. *Dumlupınar Üniversitesi Sosyal Bilimler Dergisi*, *20*, 277–297.

Elliott, G. C., Ziegler, H. L., Altman, B. M., & Scott, D. R. (1982). Understanding stigma: Dimensions of deviance and coping. *Deviant Behavior*, *3*(3), 275–300. do i:10.1080/01639625.1982.9967590

Erçetin, Ş. Ş., Açıkalın, Ş. N., & Bülbül, M. Ş. (2013a). A Multi-Dimensional approach to leadership in chaotic environments. In S. Banerjee (Ed.), *Chaos and Complexity Theory for Management* (pp. 89–104). IGI Global. doi:10.4018/978-1-4666-2509-9.ch005

Erçetin, Ş. Ş., Potas, N., Kısa, N., & Açıkalın, Ş. N. (2013b). To be on the edge of chaos with organizational intelligence and health. In S. Banerjee (Ed.), *Chaos and Complexity Theory for Management* (pp. 184–203). IGI Global. doi:10.4018/978-1-4666-2509-9.ch009

Güngör, H. (2013). *Okul yönetici ve öğretmenlerinin damgalama eğilimleri ve örgüt kültürü arasındaki ilişki. Yayınlanmamış yüksek lisans tezi*. Sakarya Üniversitesi Eğitim Bilimleri Enstitüsü.

Güngör, H. (2014). Adolescence suicide as a chaotic phenomenon. In Ş. Ş. Erçetin & S. Banerjee (Eds.), *Chaos and Complexity in World Politics*. Hershey, PA: IGI Global.

Jaspal, R. (2011). Caste, social stigma and identity processes. *Psychology and Developing Societies*, *23*(1), 27–62. doi:10.1177/097133361002300102

Jones, E., Farina, A., Hastorf, A., Markus, H., Miller, D., & Scott, R. (1984). *Social stigma: The psychology of marked to relationships*. New York: W.H. Freeman & Co.

Karasar, N. (2006). *Bilimsel araştırma yöntemi*. Ankara: Nobel Yayın Dağıtım.

Kocabaşoğlu, N., & Aliustaoğlu, S. (2003). *Stigmatisation. New Symposium*, *41*(4), 190–192.

Link, B. G., & Phelan, J. C. (2001). Conceptualizing stigma. *Annual Review of Sociology*, *27*(1), 363–385. doi:10.1146/annurev.soc.27.1.363

Major, B., & O'Brien, L. T. (2005). The social psychology of stigma. *Annual Review of Psychology*, *56*(1), 393–421. doi:10.1146/annurev.psych.56.091103.070137 PMID:15709941

Oran, N. T., & Şenuzun, F. (2008). Toplumda kırılması gereken bir zincir: HIV/AIDS stigması ve baş etme stratejileri. *Uluslararası İnsan Bilimleri Dergisi*, *5*(1), 1–16.

Parker, R., & Aggleton, P. (2003). HIV and AIDS-related stigma and discrimination: A conceptual framework and implications for action. *Social Science & Medicine*, *57*(1), 13–24. doi:10.1016/S0277-9536(02)00304-0 PMID:12753813

Şimşek, Y. (2003). *Okul müdürlerinin iletişim becerileri ile örgüt kültürü arasındaki ilişki-Eskişehir ili örneği. Yayımlanmamış doktora tezi*. Eskişehir: Anadolu Üniversitesi, Eğitim Bilimleri Enstitüsü.

Stigma ve Epilepsi. (2013). Retrieved from www.turkepilepsi.org.tr/userfiles/file/.../stigma_ve_epilepsi.pdf

Üçok, A. (2003). Şizofreni hastası neden damgalanır? *Klinik Psikiyatri*, *1*, 3–8.

Varas-Díaz, N., Neilands, T. B., Malavé Rivera, S., & Betancourt, E. (2010). Religion and HIV/AIDS stigma: Implications for health professionals in Puerto Rico. *Global Public Health: An International Journal for Research, Policy and Practice*, 1–18. Retrieved from http://www.ncbi.nlm.nih.gov/pmc/articles/PMC2891884/ PMID:20087809

Yaman, E., & Güngör, H. (2013). Damgalama (stigma) ölçeği geçerlik ve güvenirlik çalışması. *Değerler Eğitimi Dergisi*, *25*(11), 251–270.

Zorlu, M & Çalım, S. İ. (2012). İşyerinde damgalama ve ayrımcılık faktörü olarak HIV/AIDS. *Çalışma ve Toplum Dergisi, 4*, 165–188.

ADDITIONAL READING

Banerjee, S., Erçetin, Ş. Ş., & Tekin, A. (2014). *Chaos Theory in World Politics (Vol-1)*. New York: Springer Understanding Complex Systems Series. doi:10.1007/978-94-017-8691-1

Erçetin, Ş. Ş., & Banerjee, S. (2014). *Chaos and Complexity Theory in World Politics*. IGI Global, USA. http://www.igi-global.com/publish/call-for-papers/call-details/1045

Erçetin, Ş. Ş., & Banerjee, S. (2015). *Chaos, Complexity and Leadership 2013*. New York, USA: Springer Proceedings in Complexity. doi:10.1007/978-3-319-09710-7

Goffman, E. (1963). *Stigma: Notes on the management of spoiled identity*. Harmondsworth, Middlesex: Penguin Books.

Güngör, H. (2013). *Okul yönetici ve öğretmenlerinin damgalama eğilimleri ve örgüt kültürü arasındaki ilişki. Yayınlanmamış yüksek lisans tezi*. Sakarya Üniversitesi Eğitim Bilimleri Enstitüsü.

KEY TERMS AND DEFINITIONS

Chaos: Chaos can be defined as individual's getting out of the daily routine in spiritual, mental, physical, emotional or existing in another dimension of life, and also domination of uncertainty, instability to individual, group, society or the world.

Principal: Managers who administrate any school which are preschool, primary, school, secondary school, high school.

Stigma: Stigma defined as a concept that ostracized individuals and/or groups, which have unaccepted characteristics by society. Stigma is a phenomenon that based on prejudices, stereotypic thoughts and labeling. Although stigma is a phenomenon that has been experiencing all areas of life and all sectors in society and also there has been increasing research on stigma during last twenty years, studies on principals' stigma tendency is extremely limited. School principals are the most affecting actors more positively or negatively to educational organizations and also they play a key role in schools.

Stigma Scale: Stigma Scale which was developed by Yaman and Güngör (2013) and personal information form which was prepared by researchers were used for data collection. Five Likert type scale was used (1 - strongly disagree, 2 - Disagree, 3 - Partially agree, 4 - agree, 5 - Completely Agree) in the scale. The lowest score that can be obtained from the scale is 22 the highest score is 110. According to Stigma Scale evaluation the stigma tendency of individuals who get below 55 point is low and the stigma tendency of individuals who get above 55 point is high (Yaman and Güngör, 2013). Stigma scale is composed of 22 items and four subscales. Subscales are named as discrimination and exclusion, labeling, psychological health and prejudice.

Stigma Tendency of Principals: Although stigma is a phenomenon that has been experiencing all areas of life and all sectors in society and also there has been increasing research on stigma during last twenty years, studies on principals' stigma tendency is extremely limited. School principals are the most affecting actors more positively or negatively to educational organizations and also they play a key role in schools.

Chapter 7
Managing the Chaos and Complexities of Informal Organizations for the Effectiveness of Schools as Formal Organizations

Sa'adu Isa Bashar
Sokoto State University, Nigeria

Zayyanu Sambo
Shehu Shagari College of Education, Nigeria

ABSTRACT

Informal organizations in education are those groups or unions that are spontaneously formed within the educational institutions by the members of staff, students and tradesmen for the realization of their identified goals. Those groups, in spite of their advantages, are chaotic and complex in nature as they are bound to bringing about frustration to the top or line mangers of the schools, encourage negativism, generate inter-personal and group rivalries, resist changes, cause harassment to certain employees, promote rumor mongering and promote dissatisfaction among employees. All of these deter the smooth operation and management of schools which if care is not properly taken may result to institutional entropy. The chapter puts forward some future trends for managing complexity of informal organizations by contending that, educational managers should employ approaches of cooptation, divide and rule, control of information, control of reward, displacement and control of meeting for the management of those groups.

DOI: 10.4018/978-1-5225-0460-3.ch007

INTRODUCTION

There never exist variables of informal organizations except in the existence of formal organizations where human and material resources relate and work interdependently for goals. This is why Sharma (2003) defined organization as both a structure and a process. As a structure, it is the structure of human relationships which include both formal and informal that affects the behavior of employees. Hence, it is a systematic combination of people, functions and facilities. It is the framework within which people work and live together for the achievement of predetermined goals. As a process, organization involves a process of establishing harmonious relationship between departments, activities and human resources which defines the functions of all employees of an organization, decides their authorities and establishes relationship among them. Saleemi (2011) also sees organization as a structure of relationships among Individuals. It is basically a group of people formed to accomplish some common objectives.

Schools as Formal Organizations

Formal organizations have been taken to mean organizations or groups that represent the structure of duties, responsibilities and working relationships (Kawesi, 2014). Sociologists also perceived formal organizations as social systems that are organized around specific goals and usually consisting of several interrelated groups or subsystems.

Educational institutions otherwise referred to as schools, colleges or universities are places for learning and education. They are the temples of learning where learners assemble; sit together under the same roof, on the same floor, carpet or desks and learn from the same teachers irrespective of color or creed. It is in them that teaching and learning had been institutionalized (Sidhu, 1999). Educational institutions as formal organizations have the following formal characteristics that distinguish them from the informal organizations:

- They are formed by Authority. Educational institutions are formed and guided by the high level authorities like the Ministries of Education (Federal, State or Local Education Authorities), legislatures and or executives and have legal backings.
- They have determined objectives and policies. That is to say they aim at achieving target and specific objectives. This forces them towards establishing relevant policies and programs leading to the attainment of those goals.

- They have shared duties and Responsibilities. In schools, duties, responsibilities, authority and accountability of each member is well-defined. For instance, Principals and their deputies (vice principals of administration and academics) in secondary schools, are to perform administrative or managerial functions which are distinct from that of teaching – the responsibility of teaching staff. However, students have their own duties and that is abiding by the school rules and regulations as well as concentrating fully to their studies. Moreover, non academic staff have their own responsibilities of ensuring the neatness of the school, security of the school, examinations and record keeping practices are dully discharged. Every individual in the school is therefore assigned with specific duties & responsibilities to perform for the realization of the desired goals of the school.
- They have organizational Charts. Institutions of learning have organizational charts that simply portray the structure of those institutions. Those are otherwise called School or Institutional charts.
- They have well defined rules and regulations. Schools have a number of stipulated rules and regulations. The members of the school (staff and students) must therefore have to follow those rules and regulations. Those rules may be in written or verbal and are often formed by the top level authority of the institutions, constitutional provisions or legislative acts. If the members of those institutions follow those rules, they may be rewarded and failure to comply may attract punishment.
- Members strictly observe the principle of co-ordination. Educational institutions always possess certain arbitrary structures that allow all members of the faculties or departments to work in a coordinated way to achieve the targets of the institutions. This co-ordination between different departments in schools and colleges or universities is strictly maintained to achieve the most efficient result.
- They have vertical chain of communication. Normally in a formal organization like schools, vertical communication chain is followed so that the chain of command & its unity is maintained. Information and or command come from top to bottom in a vertically unidirectional manner.
- They have status symbols. Educational institutions as formal organizations are social systems which provide an outlook of the wider societies. They are therefore replica of their larger communities through which their brightness or otherwise could be determined.
- They are stable. Schools are stable in nature having a very fixed norms and values.

- They are hierarchical having Superior-Subordinate Relationship. In schools, there exists a relationship between the superior officers and the lower level staff who harmoniously work to achieve the school goals and objectives.
- They offer benefits for members. Members of staff of various educational institutions acquire some benefits like financial and job security benefits. For instance, staff benefit from wages or salaries, bonus, travelling allowances, health insurance, loans, and many more.

Informal Organizations in Educational Institutions

Informal organizations are those unions or interrelated groups that are spontaneously formed by members of formal organizations within the formal organizations for the attainment of their predetermined goals. Saleemi (2011) contends that an informal organization is the relationships between people in an organization based not on procedures and regulation but on personal attitudes, prejudices, likes and dislikes. Informal organization is a natural or spontaneous network of relationships based on personal needs, attitudes and emotions. It consists of small groups which often overlap each other as one person may be a member of many groups. Informal organization is a joint personal activity without conscious common purpose though contributing to joint result.

Informal organizations are characterized with the following features as identified by Wikipedia, the free encyclopedia (2014):

- They evolve constantly and spontaneously.
- They are dynamic and responsive.
- Treat members or individuals equally.
- They are flat and fluid.
- They are covered by trust and reciprocity.
- They are difficult to pin down.
- They make decisions collectively.

Although Informal organizations exist in lower level educational institutions, yet, they are more often found in complex educational institutions of higher learning like Universities, Colleges of Education and Polytechnics where there are a complex structure and diverse sets of interests and interest groups. Interest groups vary from one institution to another having different goals depending on the environment and the ground upon which they were initiated. In Universities for instance, the informal organizations include; Students' Unions, like Students Union Government (SUG) - who fight for students' rights, Academic Staff Union of the University (ASUU), Colleges of Education Academic Staff Unions (COEASU) and Academic Staff

Union of Polytechnics (ASUP) – who fight for freedom and rights of the teaching staff, Non Academic Staff Union of the University (NASUU) – who fight for the non teaching staff, Muslim Students' Society of Nigeria (MSSN), Associations of Christian Students (ACS), Faculty Unions (FUs), Departmental Unions (DUs) and alike. These and other unions also exist in larger secondary schools like the Federal Government Colleges and Unity Secondary Schools. All these are clear examples from the context of Nigerian educational institutions.

Purpose of Informal Organizations

Informal organizations are formed on the ground of the following number of reasons as outlined by Saleemi, (2011):

- Socialization. This is the desire on the part of the staff to socialize with fellow workmen not only to escape from the boredom and fatigue caused by repetitive nature of their specialized jobs but, also to satisfy the urge to belong and to associate with others rather than to lead a lonely existence.
- To overcome tension. This is the desire on the part of the workmen to overcome the tension caused by performance of repetitive jobs, and to experience a sense of accomplishment denied to them by the jobs as such.
- To express their abilities. This is the desire on the part of the workmen to give expression to their capabilities and talents, which are not fully used in the performance of the specialized jobs.
- To acquire information. This is the desire on the part of the staff to seek escape routes from the rigid information and communication network of the formal organization structure which often denies them messages which are meaningful from their point of view.
- To fight for their rights. This is the desire on the part of the workmen to struggle and fight for their rights that may be denied to them by their formal organizations.

Chaotic and Complex Nature of Informal Organizations in Educational Institutions

In spite of the fact that informal organizations are formed on the ground of accomplishing certain number of goals to their members, they pose a considerable number of challenges in the schools they exist which often deter the smooth operation and effective management of those institutions. Saleemi (2011) emphasized the following to be some of the complexities of informal organizations in the midst of the formal ones:

- They cause frustration to the management. This is by bringing about hindrances or criticisms as regards to the organizational plans, goals or efforts. They may result to disobedience, strikes and revolts. For instance, very recently an Academic Staff Union of Nigerian Universities (ASUU) called off strike action that caused all members of the union who shoulder the responsibility of all academic activities of their institutions to have gone for strike when it was a period in which semester exams were to be conducted. This has frustrated the management of many universities seeing that the overall academic calendar of the various institutions had to be changed to meet up the desired sessional arrangement. Similarly, in Shehu Shagari College of Education Sokoto for instance, in the year 2009, Students' Union Government revolted setting fire on many of the institution's properties because the management has failed to pay them their monthly allowances given by their respective state governments and at the same time failed to provide adequate water supply throughout their hostels. This necessitated the college to have paid them their dues and embarked on supplying adequate water to them.
- They bring about lack of motivation to members of staff. Members of those informal unions often feel discouraged or demotivated in carrying out their responsibilities or duties in the school. This happens especially in schools where employee are not fully respected, rewarded or entertained by their managers. So, as long as employees are ill-motivated, a school will no longer be able to attain any success or progress.
- They encourage negativism. Some informal unions promote negativism among the members of the school or institution. This is because they always tend to be doing things that may be contrary to the wills of the larger formal organization just in their attempt to see to the realization of their own personal target goals. This is why Tripati and Reddy (2008) contended that it was among the bad aspects of informal organizations doing what is called sub-optimization. This is a situation whereby group members put their own needs ahead of those of the organization or the group objectives conflict with or take precedence over the larger organizational goals and consequently this makes the latter suffer.
- They encourage inter-personal and group rivalries. This is also referred to as inter-personal and group disputes or arguments. Members of informal unions sometimes contribute to unnecessary disputes or arguments between their group and that of others or even among themselves, but, this depends on their goal, time, situation and organization. However, disagreement in an organization is a problem that hinders the smooth and effective operation of the school because it is the brain behind any sort of disunity with which all things of the school collapse.

- They oppose a change of any kind. The informal organizations are of the habit of resisting to school or Institution's changes especially those changes that the members perceived to detract them from their own informal cultures. This also comes when the members of those unions understand that the changes are of the aim of threatening the continuity of their existing relationship as a group. This goes in handy with what Tripati and Reddy (2008) called as resistance to change.
- They cause an undue harassment to certain employee. Informal organizations also make some fellow working personnel who may not cooperate with them feel tormented by certain continued and persistent attacks, molestations and criticisms so as to make them feel tortured until they join them. For instance, if an academic staff that did not join the ASUU is deprived from being paid his full salary or fringe benefits in his or her university, the union will rather make jest of him instead of helping him to get rid of the problem.
- They promote grapevine or rumor mongering. Since one of the basic rationales behind the establishment of informal unions is to generate information and communication that might be denied to them by their formal organizations, one unfortunate thing behind this is that they also use this medium to segregate rumors that may not necessarily have authentic root. This therefore contributes to sharing of lies and uncertainties which may even bring about another problem within the organization.
- They promote dissatisfaction among employees. This comes especially when members of the group tried to show a kind of joy or satisfaction that they might have acquired through the union like provision of loans, information, facilities and voluntary incentives. This would make other employees who have not yet joined the union to feel dissatisfied and feel uncomfortable about their employer who could not provide those things to them.
- Group-thinking Philosophy. Informal organizations as opined by Tripati and Reddy (2008) promote what is otherwise called group-thinking Philosophy in a formal organization. This simply implies a situation whereby members of the group begin to worship the group by believing that whatever the group decides is an optimum good and right whether it is wrong. This thus develops a sense of delusion of righteousness among the group members.

CONCLUSION

The chapter would not be concluded without mentioning that managing informal organizations within the context of formal organizations is a paramount necessity that requires an educational manager to pay serious attention to. This is because they enable one to have a better focus so that the organization's objectives are achieved with relative ease. It is worthy of noting here that an educational manager should not see informal organizations as a threat rather an opportunity through which school's effectiveness could be established as they point out the areas of strength and weaknesses. Those unions should be well managed following the aforementioned management strategies so that the schools could benefit from their positive sectors and at the same time be far from their negative trends. With this, a school or institution would be able to attain success by achieving their target goals.

FUTURE TRENDS FOR MANAGING THE CHAOS AND COMPLEXITIES OF INFORMAL ORGANIZATIONS IN EDUCATIONAL INSTITUTIONS

Since informal organizations are spontaneously formed and that they have strict influence on the formal issues of the educational institutions and that they could not be eliminated by school managers, the only best way to bringing about smooth operation of those institutions is to ensure that the informal organizations are well managed vis-à-vis the formal schools organizations. To achieve this, the paper suggests that the educational managers should employ the following approaches towards that direction:

- **Cooptation:** This is winning the minds or acquiring the cooperation of the leaders of those informal groups. This is because if this is not done, the leaders may at one time or the other not gives their full support to the school heads who do not carry the informal leaders with them.
- **Divide and Rule:** This is a method used by leaders to maintain power in weakly-institutionalized polities while simultaneously pursuing policies costly to society. The logic of the divide-and-rule strategy is to enable school managers to bribe politically pivotal groups of the equilibrium path, ensuring that they can remain in power against challenges.
- **Control of Information:** Here it means in most situations organizations' heads are the recipient of information. At this point they may use their positional power to influence decisions within the organization through controlling any meaningful information coming into organization. Therefore they may try to ensure that only those issues that may not result into chaos and misunderstanding in the organization are revealed to leaders and or members of the existing informal organizations in the schools.

- **Control of Reward:** Power is likely to be possessed to a significant degree by individuals who have control of rewards. In education, rewards may include promotion, good references and allocation to favored classes or groups. School managers who control or influence the allocation of these benefits may be able to determine the behavior of teachers who seek one or more of the rewards.
- **Displacement:** This strategy involves any legal procedure that is used by the head of an organization to ensure that apparent issues are used to obscure real issues in an organization.
- **Control of Meeting:** As the name suggests, this strategy calls for school managers to use their power to direct, regulate and determine the agenda of the staff meeting so that the informal union members may not be given much time to discuss their allegations in the meeting. For instance, if the school manager come to understand that members of the informal groups of his/her school may wish to agitate on the rate of salary or pay and it has been mentioned as number one point of the agenda, he/she may decide to change the agenda by delaying that issue and deliberate probably on other issues until towards the end of the meeting then the point of salary be discussed. This would deny the informal group the ability to deliberate or agitate much on that issue for the sake of inadequacy of time.

REFERENCES

Alfred, M. (2003). Kleptocracy and divide and rule: A model of personal rule. Paper presented at the Marshal European economic association annual lecture, Stockholm, Sweden.

Bernard, C. (1938). *The functions of the executive*. Cambridge, MA: Harvard University press.

Definition of formal organization. (n.d.). Retrieved on 21st February, 2014 from www.businessdictionary.com/definition/formal-organisation.html

Gaurav, A., Mudit, K., et al. (2011). Distinction between formal and informal organisation. *Kalyan City Life*.

Gibson, J. L., & Ivancevich, J. M. et al.. (2006). *Organizations: Behaviour, structure, processes* (12th ed.). New York: McGraw-Hill, Irwin.

Informal Organization. (n.d.). In *Wikipedia, the free encyclopedia*. Retrieved on 4th November, 2014 from en.m.wikipedia.org/wiki/Informal_organization

Kawesi, M. (2014). *School as formal organization*. Unpublished Manuscript, Department of Educational Management and Administration, Faculty of Education, Islamic University in Uganda.

Neeraj, K. (2006). *Principles of management*. Jaipur, India: Sublime Publications.

Saleemi, N. A. (2011). *Principles and practice of management simplified* (Revised Edition). Nairobi, Kenya: Saleemi Publications Limited.

Sharma, M. (2003). *Theory of public administration*. New Delhi, India: Anmol Publications Private Limited..

Tripati, P. C., & Reddy, P. N. (2008). *Principles of management (4th ed.)*. New Delhi, India: Tata McGraw-Hill Education Private Limited.

Chapter 8
Education and Leadership in Social Networks

Kenan İli
Hacettepe University, Turkey

Mehmet Menteşe
Hacettepe University, Turkey

ABSTRACT

Social networks which play an important role in individuals', many of whom are youngsters, daily life got the opportunity to be used as educational tools and seem to have many features to improve educational organizations and the educational processes. Organizations exist in the very complex and chaotic environments rather than the stable and steady environments. This affects the role of administrator and especially the leader within the organization. The task of the 21st century educators is to be compromising social media leadership with media, mass media instruments and information technologies and the values of the school, and preparing the youngsters for the future by building common living environments. Within the scope of research, some explanations from literature about the concept of social networks and the use of social networks in education were given first. For conclusion, comments, conclusions and suggestions about the usability of social networks in education were presented.

INTRODUCTION

Very important social changes occur in the information society we are living in owing to rapidly evolving communication technologies. With the development of new communication technologies, the concept of social media has penetrated the

DOI: 10.4018/978-1-5225-0460-3.ch008

lives of a large part of the community. Whether young or adult, everyone's profile is available on social networks, which is the most important product of social media, and they meet their daily communication and social needs over these networks (İli & Erçetin, 2014).

Information technologies reconfigured the learning-teaching process as well as the social, economic and cultural life in 21st century called information and technology age. In accordance with these evolutions, technologies applied to meet the individuals' educational needs led to changes in the educational needs, and pulled the education through its traditional school and classroom based formal structure, causing it to focus on social and informal learning (Ekici, 2012).

On the other hand, this swift evolution brought about the changes in individuals' roles and led the expected roles of educationals to complexify and diversify gradually. The more the expectation from the schools to form more efficient and sufficient learning groups increased the more the comprehension of leadership roles of the educationals in employing and practicing technology gained importance (Afshari, 2009).

As the social technologies changed the relationship between the persons, they required new leadership approaches and skills for school administrators as well. Now that, school administrators should give up controlling everything and leave the responsibility to the other administrators and teachers at the schools. In this way, leader can focus on the most difficult strategic decisions which are worth a leader's attention rather than paying attention on each decision to be made (Li, 2010).

Thereby, the use of social networks in education and the relationship of the leadership in social networks with education and social network culture were focused in this paper. Within the scope of research, some explanations from the literature about the concept of social networks and the use of social networks in education were given first. For the conclusion, the comments, conclusions and suggestions about the usability of social networks in education were presented.

Social Networks and Education

After the emergence of the internet, there happened to be many changes in the ways of communication and the daily life experiences of the persons; and the social networking in the social media is in the center of that emergent ways of communication in the virtual platforms. In those networks, people, profiling themselves, communicate with the others who are at the same cultural level and easy to get along via the opportunities of the internet (Murray, 2008). Social network can be defined as an online service, platform or domain where the people can make a social contact, communication and sharings. In this context, social networks are the theoretical structures in which the relationships between the individuals, organizations and even the communities can be observed in social sciences (Erçetin, 2013a). Although their categories and aims of usage are different, people might conduct their real social life communication, interaction and even learning activities through these social networks (Tiryakioğlu & Erzurum, 2011).

Use of social networks enriched by the opportunities provided to users in recent years is becoming more widespread every day. Given the fact that the large part of social networks users are comprised of students, teachers and school administrators, the existence of ideas and works for the educational usage of social networking sites is required to stay abreast of the changing information technologies. On the other hand, due to the dense number of social networking users, the importance of their usage by educators both for professional development and educational context is increasingly understood (Grant, 2008).

The rapid development in information and communication technologies took effect in the field of education and brought about changes in the structure, practice and the scope of the education along with the roles of the people in this context. A number of education institutions have been in search of new models to meet the needs of contemporary students. As a result of this point of view, social networks which play an important role in individuals', many of whom are youngsters, daily life could settle in the educational context with its own opportunities and those instruments could be used as educational tools. Considering from this point of view, social networks have a number of features to improve educational processes. One of the main features of social network assisted education is to increase the student-teacher interactions of traditional education systems, and gather up and use the digital tools. By the social network assisted education environment, an alternative teaching and learning which puts audio and image and transcript and image together as in the video talk on the internet can be available. Also, by means of social networks, teachers' and students' forming up an education aimed group by following some simple steps, sharing, communicating and getting feedback are very easy. In addition to these, social network sites provide the teachers with a support in teaching and evaluation processes and students with a support in learning process by enriching the learning and teaching processes with the materials such as texts, videos and audios (Togay et al., 2013).

The results of a research by Togay et al. (2013) confirmed that supporting the teaching process by social media is effectual and facilitative in learning, and makes the students feel they are more active in learning and parts of the learning process. These results revealed that employing social media tools to improve education processes can be effective and useful.

The teachers and the headmasters, in the study done by Menteşe (2012) about the use of social media environment and instruments in education, expressed their opinion that they increase the level of the education, strengthen motivation, make students more social, however, employing those instruments in all lessons with all levels of students is not appropriate.

Öztürk and Akgün (2012) also are of the opinion that social networks have the potential of supporting more efficient and sufficient learning. They stated that especially the university students are more willing to apply social networks in education process formally than the other groups of students. Munoz and Towner (2009) suggested that the subject related pictures, messages, assignments, cases and applications are to be shared on Facebook, social sharing site, for learning.

Genç (2010), in his research, emphasized that the technologies after Web 2.0 offered alternative learning environments to the traditional in-class learning, and the students who participated in this survey have a quite positive opinion about those environments at the same time, Facebook is regarded as a learning tool as well as being a social communication tool.

A study conducted with college students by Churchill (2009) revealed that a blog-based environment can help foster a learning community in which learners feel they are an important part of the classroom and that their needs and opinions matter. Findings also showed that blogs are most effective when they are designed to facilitate student access of course material, posting reflections on artifacts created through the learning tasks, and commenting on peer contributions.

The number of the studies about the functionalities of Web 2.0 tools, with a great deal of learning and teaching related features, and their extensions, social networks, in educational settings is scarce (Ajan & Hartshorne, 2008). The studies are usually on the structure of social networks, confidentiality and identity management. The summary of the educational results of these studies in the field are presented below.

As pedagogical tools, the potential benefits of social networks in educational context are social support and connectedness, collaborative knowledge discovery and sharing, forming the content, clustering of knowledge and information and modification of content (McLoughlin ve Lee, 2007). Muijs et al. (2010) claim that the social networks in educational context have the benefits of developing the school, enhancing the opportunities and source sharing, as well.

Ajan and Harsthone (2008) found out that the tools like social networks promote students' learning, supply school-student and student-student interaction, increase the lesson satisfaction of the students, improve writing skills and are the integration friendly tools.

Karaman and et al. (2008), in their literature search called "Studies Related to the Use of Web 2.0 Applications in Education", ascertained that blogs and wikis have been observed mostly, however, the number of the publications many of which centering on the primary and undergraduate levels have been gradually increasing.

Also, it is apparent in all the results that social networks have positive consequences in terms of student motivation, effectiveness and communication. Albion (2008) claimed that the more the technology advances swiftly the more the importance of the use of these instruments increases and it is critical that the educators should have the opportunity to use technology effectively and they have to be prepared for this process.

Leadership in Social Networks

Although there are various definitions of leadership by different writers and researchers in particular fields of study from different aspects, it is a fact that there is not an agreed, certain definition. Erçetin (2000) defines leader as the one who can catch the swift change in the globalizing world and manages by regarding cultural diversities and differences while the leadership is the process in which the features such as determination, risk taking, self confidence, paying attention on ethical values, being able-minded are prominent.

Leadership is an interaction domain between the leader and the followers. Because the leadership is perceived as the integration of the leader with the followers. Thus, both the leader and the followers support and take part in leadership. Since the leader does not consider himself separate from the followers, he does not ignore or deny their contribution, participation and needs. The fact that everybody can be a leader or play a leadership role is one of the significant changes in leadership perception of 21st century.

This change means the acceptance of the fact that the leadership effect does not have permanentness. Therefore, the leaders aware of the impermanency of the leadership have to create the occasions to better understand themselves and the followers. In other words, they should share leadership. As a result of this sharing, leaders take advantage of the potential of their followers at an utmost level and become more facilitator and effective in adopting the managerial decisions (Erçetin & Kamacı, 2008).

Organizations exist in the very complex and chaotic environments rather than the stable and steady environments. This affects the role of administrator and especially the leader within the organization. Accordingly, leaders should hold the organization at the edge of the chaos and demonstrate appropriate leadership behaviors to manage the change, keep the system flexible and dynamic, provide variety and compete. Leader's role is to interact with everything going in and especially around the organization and help the organization to create its own identity (Bloch, 2005). Removal of the borders in a variety of areas with the improving technology and new employment opportunities of information technologies require novel leadership strategies. The uncertainty and chaos process emerged after this wind of change brought about a new perception to the leaders and the people they lead (Erçetin et al., 2013b). We cannot think the education system apart from the chaotic, complex environments and uncertainties of today's. Coping with the uncertainties with a well-balanced management will help the system go healthy and reach the goals of existence. Well-balanced management means the shared aims, visions and decisions of every member of the system. The unpredictability is again an opportunity for the educational system to be intelligent, ready and steady to renew itself, with doubt and anxiety at a level to be self-organizing and self adaptive. The educational system can renew itself with the management, being more decentralized and knowledge centralized (Kayman, 2014). According to the chaos and complexity theory, the main role of administers are to form and create the environment where the self organization can realize appropriately. Regarding the fact that almost everything changes constantly in the current information society, it is impossible to claim that educational organizations have the necessary information to make plans for the future. At this point, what is significant for educational organizations is to activate new structuring to solve unexpected problems rather than being managed by a mechanical approach. The members of the educational organizations should adopt continual reorganizing styles together with team work and project squad. Constant refreshment of teams may mean the individual's creativity and the full use of the potential. Likewise, the purpose of the educational administrators should not be maintaining the order. Such an understanding with self-repeating can create a barrier to creativity and transformation. In case of complex and unexpected circumstances, educational organizations are to reorganize themselves and look for new orders. The only way to achieve this is to provide the system with a constant flow of new information and energy. Also the fact that small changes can have butterfly effect is to be kept in mind (Çobanoğlu, 2008).

This novel perception necessitated new leaders who are able to meet the expectations. Future leaders are thought to be thinking globally, yielding with the improving technology and sharing the leadership (Goldsmith et al., 2003). In this context, social network leadership gains importance. Social network leaders, as a leadership feature, work up various connection networks based on trust with the people working around and adopt employing social network instruments effectively in the realization of the aims process as a principle. The followers may not use social technologies as they do not feel the need of them. So, the duty of the modern day leaders is to divert staff's minds and behaviors, inspire them to use social network technologies, and train the followers in gaining the use of social network technologies skill (Li, 2010).

Recently increasing in prevalence, social networks, which have the potential of orientation nearly in every aspects, thus, are to be managed well in the century, confront us as an ever growing power. It is vital for future that this power should be canalized into the education systems and is ensured to be used effectively. Considering the fact that the users of social networks are commonly youngsters, the ideas and the studies aimed at the educational use of the social sharing sites are necessary for keeping up with the improving information technologies. On the other hand, since there are a great number of social network users, the importance of the educators' using those tools in both professional and educational contexts is understood (Özmen et al., 2011).

Skyrocketed network society in the globalizing world enabled the borders to be removed and naturally innovations to be made. It can be claimed that, in the future, the administration of the educational institutions will be maintained in the light of network society and the institutions deprived of this light will try hard to find their ways in the dark (Dönmez & Sincar, 2008). Hence, in an ever changing environment of information and technology, future educators should develop a self perpetuating perception. Educators who cannot be indifferent to the limitless information case produced by technology should see the change and gain a new identity themselves. The increase of expectation appropriate for the future needs obliges to think broadly especially in education.

The swift change in the information technologies made new model seeking for administration of various organizations unavoidable. Social network leaders, as future administrators, will have to have enough level of information and skills to manage the use of technology in order the technology used in their institutions to fit the purposes and gain maximum favor besides using technology excellently.

The fact that social networks as the products of Web 2.0 technology can be used in educational domains and probably this internet technology will be the most important teaching and administration tool is the most notable one recently. Social network leader may integrate social network technologies with the administrative processes such as knowledge sharing, communication and decision taking especially in educational organizations. At this point, a novel, social network based leadership model to manage both the educational organizations and the other organizations effectively meets us (Li, 2010).

CONCLUSION

The use of social networking is proliferating at all levels of education. It is being used to communicate with students and the community, increase collaboration and expand the learning environments outside of school walls.

With the advent of the web, the way we produce and manage knowledge has been profoundly changed, thanks to an extraordinary technological development which makes available ever more versatile instruments that are able to access and manage increasing quantities of diversified information (audio, video, writing, etc...). The web changes the way in which people and organizations learn, producing a direct impact on the nature of knowledge and the way in which people use it and transform it. In the system of knowledge that we inherited, the flow of information was slower and organized differently around entities which managed this flow (academia, public and institutional information services, etc.). The information to which we had access was limited, selected and organized, and we could dedicate time to the study of each item according to our interests and/or needs. Nowadays, the social network has obliged us to change our perspective completely, because the flow of information has increased dramatically both in terms of quantity and sources of distribution (Capogna, 2010).

As we look to the future, we can continue to see change happening in education. The combination of tight budgets, enhanced technology and the need for better educational attainment are driving this change. It will require a shift in how we teach and how we assess students and faculty alike. It may even impact how we organize our schools, how we prepare teachers and how schools deliver on their educational missions.

The teaching practices promoted by the overall experience are aimed at renewing the traditional conception of education based on a passive and notional view of teaching instead of a vision of learning that places the subject at the centre of social reality. Indeed, the educational process conveyed and conveyable through the use of these environments brings about radical change, and introduces the logic of learning in place of traditional teaching.

Leadership's real mission is to open doors of opportunity, growth and personal development for others. Old-style leaders see everything as a problem, threat or risk. This negativity infects the people they lead, who become pessimistic and fearful. Future cannot be estimated reviewing the past. With the improvements of social technologies, a serious paradigm shift in education is needed. Thus, the idea, lifestyle and types of behaviors that we try to adopt next generations at the schools cannot be isolated from the aforementioned environmental effects. The ideal is to reconcile social network leadership with media, mass media, information technologies and school's values and provide common living space. For this reason, school should be reformed and redesigned. In this context, educational constituents such as class, technical and administrative tools and teacher's role need to be reconsidered.

Now that, together with the increase of the internet use, individuals transmit their knowledge, experience and even the skills each other through social sharing environment. Therefore, instead of the traditional education perception which includes face to face information transfer in special buildings, an education environment perception which includes the use of different information and communication technology tools and regards the world as a school or even a class is to be adopted.

Since the scientific and technological improvements influence the education process, the change is too rapid. The increase of the responsibility in education causes the increase in administrators' demand in guiding their schools and the teachers' demand in orientating their classes. Management, in its simplest form, should be related with the start of the changes and implementation at schools in order for the students to improve learning. The very essential task of the school administrators is to make a difference about the improvement in their schools by leading the students, teachers and society.

Is it possible to take the advantage of this complexity? What may be the outcomes of this? Reviewing the structure of the organization in the experienced chaos and complexity, learning how to manage the organization, seeing the importance and major impacts of small changes, living together by regarding the chaos and complexity as natural mechanisms will provide exceptional opportunities for today's organizations in understanding and managing the change. The important thing is not to fall into chaos, but to be able to benefit from the complexity because the confusion will bring a new order of structure together. Chaos theory can be an important opportunity to make the necessary arrangements in education.

Especially as a result of developments in technology, positive changes and developments in education were also provided but the schools keeping the traditional leadership models which are very common in Turkey stayed away from moving with the times. Educational administrators have prominent roles and responsibilities in performing and improving the education at schools. Especially, National Education Ministry may take a facilitative role in carrying out the activities such as in service trainings and team works for the use of social networks with educational purposes by administrators and other staff in educational organizations.

Besides all these, the tasks to be discussed are evident: Which education level is more appropriate for starting the social network based education? How much does the use of social networks in education cost, and how can this be sponsored? Is an additional organization necessary for the use of social networks in education? How should the school administrators and teachers be trained in the social networks? Does the use of social networks in education necessitate changes in curriculum, teaching methods and course books and if it does, what are they?

RECOMMENDATIONS

- Technical infrastructure necessary for the effective use of social media environment and tools in education at every level of education should be set and teachers should be supported more.
- Educational administrators and teachers should be motivated to use social networks for professional development and knowledge sharing.

- Educational programs should be redesigned according to the technological improvements related to the use of social media environment and tools in education.
- Teachers should be stimulated to use the social media environment and tools in education.
- Educational administrators should convert schools into an interactive structure.
- Making the educational programs and domain a part of Web 2.0 tools will be beneficial for both students and teachers.

REFERENCES

Afshari, M., Bakar, K. A., Luan, W. S., Samah, B. A., & Fooi, F. S. (2009). Technology and school leadership. *Technology, Pedagogy and Education*, *18*(2), 235–248. doi:10.1080/14759390902992527

Ajan, H., & Hartshorne, R. (2008). Investigating faculty decisions to adopt Web 2.0 technologies: Theory and empirical tests. *The Internet and Higher Education*, *11*(2), 71–80. doi:10.1016/j.iheduc.2008.05.002

Albion, P. (2008). Web 2.0 in teacher education: Two imperatives for action. *Computers in the Schools*, *25*(3/4).

Bloch, D. P. (2005). Complexity, chaos and nonlinear dynamics: A new perspective on career development theory. *The Career Development Quarterly*, *53*(3), 194–207. doi:10.1002/j.2161-0045.2005.tb00990.x

Capogna, S. (2010). Learning and social networks: New forms of sociability. In *ESA Research Network Sociology of Culture Midterm Conference: Culture and the Making of Worlds*.

Churchill, D. (2009). Educational applications of Web 2.0: Using blogs to support teaching and learning. *British Journal of Educational Technology*, *40*(1), 179–183. doi:10.1111/j.1467-8535.2008.00865.x

Çobanoğlu, F. (2008). Değişim mantığını anlamak: Akış ve dönüşüm olarak örgüt. *Pamukkale Üniversitesi Eğitim Fakültesi Dergisi*, *1*(23), 110–119.

Dönmez, B., & Sincar, M. (2008). Avrupa birliği sürecinde yükselen ağ toplumu ve eğitim yöneticileri. *Electronic Journal of Social Sciences*, *7*(24), 1–19.

Ekici, M., & Kıyıcı, M. (2012). Sosyal Ağların Eğitim Bağlamında Kullanımı. *Uşak Üniversitesi Sosyal Bilimler Dergisi*, *5*(2), 156–167.

Erçetin, Ş. Ş. (2000). *Lider sarmalında vizyon*. Ankara: Nobel Yayınları.

Erçetin, Ş. Ş., Açıkalın, Ş. N., & Bülbül, M. S. (2013b). A multi-dimensional approach to leadership in chaotic environments. In S. Banerjee (Ed.), *Chaos and Complexity Theory For The Management: Nonlinear Dynamics* (pp. 89–104). Hershey, PA: Business Science Reference; doi:10.4018/978-1-4666-2509-9.ch005

Erçetin, Ş. Ş., & Kamacı, M. C. (2008). Quantum leadership paradigm. *World Applied Sciences Journal, 3*(6), 865–868.

Erçetin, Ş. Ş., & Neyişçi, N. B. (2013a). Social networks: Connections in structures. In Ş. Ş. Erçetin & S. Banerjee (Eds.), *Chaos, Complexity and Leadership 2013*. Springer Proceedings in Complexity. doi:10.1007/978-3-319-09710-7_26

Genç, Z. (2010). *Web 2.0 yeniliklerinin eğitimde kullanımı: Bir facebook eğitim uygulama örneği. Akademik Bilişim'10 - XII*. Akademik Bilişim Konferansı Bildirileri 10 - 12 Şubat 2010 Muğla Üniversitesi.

Goldsmith, M., Greenberg, C., Robertson, A., & Hu-Chan, M. (2003). *Global leadership - The next generation*. Financial Times Press.

Grant, N. (2008). On the Usage of Social Networking Software Technologies in Distance Learning Education. In K. McFerrin et al. (Eds.), *Proceedings of Society for Information Technology and Teacher Education. International Conference*. Chesapeake, VA: AACE.

İli, K., & Erçetin, Ş. Ş. (2014). The impact of social media environments and tools on student behaviour. *US-China Education Review B, 4*(5), 295-302.

Karaman, S., Yıldırım, S., & Kaban, A. (2008). *Öğrenme 2.0 yaygınlaşıyor: Web 2.0 uygulamalarının eğitimde kullanımına ilişkin araştırmalar ve sonuçları. XIII*. Ankara: Türkiye'de İnternet Konferansı Bildirileri, Orta Doğu Teknik Üniversitesi.

Kayman, E. A. (2014). Chaos in education as an intelligent complex adaptive system. In Ş. Ş. Erçetin & S. Banerjee (Eds.), Chaos and Complexity Theory in World Politics (pp. 289-298). IGI Global.

Li, C. (2011). *Açık liderlik: Sosyal teknoloji yönetme tarzınızı nasıl değiştirir?* İstanbul: Optimist Yayınları.

Mcloughlin, C., & Lee, M. J. (2007). Social software and participatory learning: Pedagogical choices with technology affordances in the Web 2.0 Era. In ICT: Providing Choices For Learners And Learning. Proceedings Ascilite Singapore.

Menteşe, M. (2013). *Sosyal medya ortam ve araçlarının eğitimde kullanımına ilişkin okul yöneticilerinin ve öğretmenlerin görüşleri*. Ankara: Yüksek Lisans Tezi, Hacettepe Üniversitesi.

Muijs, D., West, M., & Ainscow, M. (2010). Why network? Theoretical perspectives on networking. *School Effectiveness and School Improvement*, *21*(1), 5–26. doi:10.1080/09243450903569692

Munoz, C., & Towner, T. (2009). Opening facebook: How to use facebook in the college classroom. In Society For İnformation Technology & Teacher Education İnternational Conference.

Murray, C. (2008). Schools and social networking: Fear or education? *Synergy Perspectives: Local*, *6*(1), 8–12.

Özkan, B., & Mckenzie, B. (2008). Social networking tools for teacher education. In K. Mcferrin et al. (Eds.), *Proceedings of Society for Information Technology and Teacher Education International Conference*. Chesapeake, VA: AACE.

Özmen, F., Aküzüm, C., Sünkür, M., & Baysal, N. (2011). Sosyal ağ sitelerinin eğitsel ortamlardaki işlevselliği. In *6th International Advanced Technologies Symposium (IATS'11)*.

Öztürk, M., & Akgün, Ö. E. (2012). Üniversite öğrencilerinin sosyal paylaşım sitelerini kullanma amaçları ve bu sitelerin eğitimlerinde kullanılması ile ilgili görüşleri. *Sakarya University Journal of Education*, *2*(3), 5–27.

Tiryakioğlu, F., & Erzurum, F. (2011). Bir eğitim aracı olarak ağların kullanımı. In *2nd International Conference on New Trends in Education and Their Implications*. Retrieved from www.iconte.org

Toğay, A., Akdur, T. E., Yetişen, İ. C., & Bilici, A. (2013). *Eğitim süreçlerinde sosyal ağların kullanımı bir MYO deneyimi*. XV. Akademik Bilişim Konferansı, Akdeniz Üniversitesi, Antalya. Retrieved from http://ab.org.tr/ab13/bildiri/302.pdf

KEY TERMS AND DEFINITIONS

Chaos: In common usage, "chaos" means "a state of disorder". However, in chaos theory, the term is defined more precisely. A commonly used definition is that, for a dynamical system to be classified as chaotic, it must have the following properties; 1.it must be sensitive to initial conditions,2.it must be topologically mixing, 3. its periodic orbits must be dense.

Chaotic Situation: A complex situation which emerges as they do not know sufficiently necessary information in order to predict a situation about the future and the relationships between this information.

Complexity: The state or quality of being intricate or complex. Meanwhile in this case complexity reflects the intricate nature within the system of managing human resources in education systems; this renders them complex at all levels.

Social Network: A dedicated website or other application which enables users to communicate with each other by posting information, comments, messages, images, etc.

Chapter 9
Complexity of School Leadership in the Wake of Teacher Insularity

Ssali Muhammadi Bisaso
Islamic University in Uganda, Uganda & Hacettepe University, Turkey

ABSTRACT

The interest to all actors and observers within educational arena denotes a long-term trend of expansion of educational professionalism accompanied by a continuous debate about its desirability, and a perennial instability or dynamic of the structures within education systems. Over the years, however, the emphasis placed on shape of professionalism in education systems has varied substantially to reflect the complexity contained therein from a purely managerial and leadership perspective. Moreover, we observe considerable changes of ''views and controversies'' concerning the most desirable quality and structure of teacher interaction and professionalism in educational institutions. This chapter explores the state of teacher interaction and collaboration in educational institutions as well as dissecting implications of same for management and leadership in schools. The study is deemed an eye opener to managers, leaders in schools to 'mind the gap' that is teacher insularity if they are to effectively achieve the very best out of them.

INTRODUCTION

In today's rapidly developing and changing world, people's life and thinking, values, expectations and problems have changed and with the impact of globalization were in a process in which people influence each other even more. When teachers

DOI: 10.4018/978-1-5225-0460-3.ch009

are considered to be in the center of education and training, then the fact that one of the occupational groups affected most by those developments is the teaching profession becomes an undeniable fact. Therefore in this profession, detection of problems and seeking of solutions continue to be very important phenomena not to be neglected. Dwelling on the problems especially starting from the training process of teachers will have a positive impact on the success of education which is a long-term investment (Demir, 2013). The training should prepare teachers for the daunting and challenging task ahead of them when they become fully fledged teachers within the classroom.

Trygestad (1997) indicates that classes are non-linear systems, chaotic and having unpredictable processes. Teachers respond to chaos with classification method in order to reduce instability and undesired behaviors and to increase stability and rate of behavior. Educators usually indicate that order exists and they deny or underestimate disorder and the errors. They also acknowledge the fact that those errors and disorders are a coincidence. But errors are factors requiring investigation for proper analysis of the system. Errors that we accept as a coincidence in education include information needed for restructuring the system (Radford, 2006).

Today there are many roles and duties which are expected of teachers. These roles are important today as they were in the past and expected to maintain their relevance in the future. Radford (2006) asserts that although duties and responsibilities of teachers are basically the same and indeed form the basis of teaching, changes may occur in some responsibilities due to some details and developments emerging over time.

Indeed today's schools are more ''complex systems than the one-room schools'' of the past (Cahill, 2010; Cunnigham, 2000; Kara, 2008). However, not much has changed about beliefs, perceptions and expectations about schools today once compared with the situation in the olden days. In the one-room schools of old times, the teacher was responsible for all the instruction of all the students, the maintenance of the building, keeping the stove filled with wood and cleaning the floors (Kara, 2008; Lortie, 1975). In a one-room school, ''the teacher was responsible for all that transpired within its four walls. This legacy of independence and isolation remains alive and well in many schools today'' (Cunningham, 2000; Glickman, 2001). Although the old ''one-room school'' is physically gone, it still pervades the minds and actions of many teachers and administrators of today. Accordingly, although teachers were regarded as people transferring information to students in the past, today teaching is no longer found adequate, instead they need to participate in efforts regarding guiding, predicting future results concerning decisions, engaging social processes and protecting nature (McNeil, 1996)

Issuing from the above premise therefore, with all the pressures they face in the classroom, one might think that teachers would work together to establish strong relationships with their fellow educators. Such relationships could serve to alleviate their self doubt, re-establish self worth and rekindle teacher determination as opined by McPherson (1970). They could help to establish and reinforce the educational aims of the school and facilitate the sharing of workable methods and worthwhile ideas. However a critical and closer look at the contemporary setting of teacher-teacher interaction in schools will only present more and more insularity and superficiality that drives school leadership (management) to the edge of chaos.

Teacher insularity has been expressly explored by a host of researchers to date (Farenga, 2013; Heider, 2005; Kasloff, 2007; Lortie, 1975; Murray, 2007; Steyn, 2013). Even other authors have picked interest in writing about insularity or isolation of teachers in the school system (Maistry, 2008; McPherson, 1970; Mirel & Goldin, 2012; Wasley, 2012). In many of these works, the nature of teacher insularity in schools (Farenga, 2013; Mirel & Goldin, 2012; Wasley, 2012; Steyn, 2013), causes of teacher insularity (Ashton & Rodman, 1986; Carroll & Fulton, 2004; Heider, 2005; Hull, 2004; McPherson (1970), positive impact and negative returns of teacher insularity (Lortie, 1975), complexity of leadership in view of teacher insularity (Shyamala,) and intervention mechanisms to teacher insularity (Heider, 2005; Hirsch, Koppich & Knapp, 2001; Maistry, 2008) are described and examined effectively.

In analyzing the nature of insularity, Wasley (2012) contends that the most detrimental condition of a teacher's work life is the isolation in which the majority of teachers work. The one-teacher-in-one-classroom model that most schools still use is based on an old set of turn-of-the 20th century assumptions that focused on factories and efficiency. Wasley (2012) also contends that teaching, however, is a human enterprise and teachers need feedback from knowledgeable others to improve their own skills and to expand their ability to assess what the children in their classrooms need in order to reach higher standards.

Meanwhile Farenga (2013) investigated the insularity of school research and explored the solipsistic nature of school research in which it is claimed that educationists avoid discussing the rather contentious issues of schools and education in general. This is based on the essay by Houman Harouni titled "A question of silence": Why we don't read or write about Education. This implies that not much about education can be examined since all teachers keep key issues to themselves.

Indeed confident and capable teachers demonstrate complete dedication to their students and professional responsibilities, share pleasant social relationships both inside and outside the school, and care about each other and enjoy working in this warm environment (Kasloff, 2007). However, on a professional level, isolation is the norm. Some teachers also spend the teaching day isolated from other adults; the initial pattern of school distribution in such a case represents a series of 'cells' which are construed as self-sufficient (Lortie, 1975).

It is also argued that educators spend most of their time distanced from their colleagues (Mirel & Goldin, 2012). Indeed Mirel and Goldin (2012) assert that ''Instead of forcing them to compete with each other, we should help them find new ways to work together''. Issuing from the above views, it becomes clear that teachers feign unity and collaboration in schools but the truth of the matter is that they are in critical and detrimental isolation.

Maistry (2008) actually opines that more studies need to be done to understand the conditions that would enhance and sustain collaborative structures for teachers in schools, especially in developing countries. Accordingly, Steyn (2013) in 'Building Professional Learning Communities to Enhance Continuing Professional Development in South African Schools' contends that isolated teaching practices are counterproductive for improving teaching practices and interdependency is a requirement for teachers' collaborative learning and more effective teaching practices.

Premised on this acute call, Murray (2007) conducted a study titled countering insularity in teacher education: academic work on pre-service courses in nursing, social work and teacher education' in which a comparison of the practices of teacher educators with those of academics working as educators on pre-service courses for nurses and social workers is made. The findings illuminate generic issues about what is termed second order practice, and indicate directions for countering insularity in teacher education. The bottom line in Murray's study is the fact that teacher insularity is eating up education systems the world over.

In keeping with the need to encounter teacher insularity, Heider (2005), claims that teacher isolation (insularity) or attrition has become a very serious problem in the United States in recent years. He equally argues that many talented, new teachers are leaving the profession early in their careers due to "feelings of isolation". The article thus examines the potential of mentoring programs in reducing teacher isolation or insularity in schools. Yet as far as Lortie (1975) is concerned, "ongoing professional development programs" are also needed to support teachers' substantive work together.

Virtually, no educational institution has been devoid of teacher isolation and programs to help teachers teach in collaboration have always been launched in which it is stressed that teachers should never work in isolation. Instead it is believed that the power of collaborative teaching is as an integral approach to embed global themes within the existing school curriculum and help teachers work together.

It is also known that some teachers are uncomfortable with certain technical idiocies of pedagogy in particular and education in general yet they deal with the teaching and learning process from time to time (Bisaso & Rafiyyah, 2010). But since these insecurities are not overtly revealed, self-reliance is maintained by the teachers, yet failure to discuss some fundamental questions about the learning process only serves to guard teachers against possible criticism from colleagues (McPherson, 1970).

The most glaring undoing in the end is that teachers don't quite get feedback from colleagues in order to know what they are doing which also hinders the development of worthwhile perceptions about the learning process and leaves them fundamentally alone and apprehensive to challenges inherent within pedagogy which is central to their profession. But the biggest challenge is seen within the realm of school management and leadership. Indeed teacher isolation would make it very complex for school leaders to realize the aims and goals of the schools since collaborative teaching is key to both teacher performance and school leadership.

This paper sought to examine the complex nature of school leadership in the wake of the intolerable teacher isolation that has engulfed the schools today by exploring the causes, implications (both positive and negative), the challenges encountered by the school leaders as well as venturing into the plausible intervention mechanisms.

METHODOLOGY

The purpose of this study is threefold. First it explored the extent of teacher insularity in the schools. Second, it examined the implication of teacher insularity on teachers as well as school management. Third, it sought to unearth plausible intervention mechanisms to teacher insularity in schools. In tandem with the purpose of the study, the paper "inquired thus":

1. What are the major causes of teacher insularity in schools?
2. Of what implication is teacher insularity to schools?
3. What complexity is encountered by administrators leading under teacher insularity situations in schools?
4. Which intervention mechanisms can be engaged to curb teacher insularity in schools?

The methods employed in gathering data that enabled the writing of this paper were the literature reviewing methods of content analysis and documentary review. The approaches and techniques employed featured exploration of related web sites, paper presentations, relevant documents, as well as researches done on the field of teacher insularity. Content analysis is a research method focused on the actual content and internal features of media and/or literature (Amin, 2005).

As a way of promoting a smooth flow of the analysis, critical viewpoints were developed in line with the major units of teacher insularity. The analysis thus followed the following elements; description of teacher insularity, research on teacher insularity, the major causes of teacher insularity, the positive and negative implication of teacher insularity, complexity of managing teachers in a complex insularity setting as well as the major yet plausible forms of intervention to teacher insularity.

It is hoped that the conceptual paper shall be of critical relevance in the following ways:

- Teachers will realize the indispensable nature of "collaboration in the teaching and learning processes" as a way of transforming both students' as well as school performance.
- School administrators will be challenged to conceive mechanisms of instilling a spirit of cooperation among teachers as well as putting in place strategies to achieve the same.
- Future researchers with a bias on management, leadership or teacher insularity will use the study as a background to further research as a way of adding to the growing body of knowledge especially in the critical field of insularity.

Major Causes of Teacher Insularity in Schools

The major causes of teacher insularity in schools have been examined by a number of authors (Ashton & Rodman, 1986; Carroll & Fulton, 2004; Heider, 2005; McPherson (1970). Heider (2005) indeed contends that many of the nearly 540,000 teachers who moved to other schools or left the teaching profession in 2000 did so due to ''feelings of isolation''. In view of this, there would be need to replace the departing as well as retiring teachers which will call for new teachers to be hired to support the ever increasing enrollment in public schools (Hull, 2004). Without ''good mentoring programs'' which according to Hull (2004) can reduce teacher isolation, and encourage teachers to stay in the profession, however, it will be very a complex undertaking for schools, especially low-income schools, to fill these positions and or keep them filled at all times.

It may not be easy to map out the causes of such insularity for they may vary from one school to another but a closer look at the major ones based on authors above may reveal the following:

1. **Physical Barriers:** The setting of the schools itself which calls for a teacher locking him or herself up in the classroom does little to solve the problem anyway since it makes the teaching and learning process more of mind your own business.
2. **Social Norms:** Also play a big role in widening the gap between teachers in view of interaction since they hide in the pretext of respect for each other as well as confidence and belief in the credibility and integrity of each other.

3. **Sketchy Knowledge:** Most of the teachers either have less command of the material and would not want their colleagues to discover the trick, or the other teachers are not abreast with the concepts in another teacher's course or subject and therefore fear being exposed.
4. **Dishonesty in Self-Assessment:** Teachers never talk about the reality regarding their classroom escapades at all. Ashton and Rodman (1986: 45) argue that: "You never go into a teachers' lounge and hear a teacher say I really failed the kids today, you always hear how a teacher was on top of the situation…"
5. **Misinterpretation of Other Teachers' Reports:** Here whenever a teacher tries to report anything regarding "the teaching and learning process", they are put down and actually forced to change behavior. McPherson (1970: 151) writes that: "A teacher's report of classroom success even couched in appropriately modest language, is often seen by colleagues as self-advertisement and may not be taken seriously."
6. **Jealousy over Success:** Every teacher may not be willing to know about the other's success story and therefore to keep away from the same, do not venture into another's business and so he won't do the same to yours.

The Implication of Insularity on Teachers' Relations in the School System

Teachers are careful to avoid confrontation with one another and to keep school relationships calm and free of friction. They are slow to criticize each other and when compelled to do so, they usually keep their comments passionless and subtle. Although it may sound bad therefore, teacher insularity is equally healthy on some grounds. The *Positive Implication* of the teacher insularity laden situations will therefore be as follows;

- School peace based on the ideology of noninterference within another's business which implies less or no conflicts in the school system.
- Self-confidence will develop within the teacher as they will feel a sense of independence as well as self-reliance.
- A feeling of authority and protection of classroom autonomy will be born and this will be key to raising a teacher's self-esteem in the school system.
- Levels of school solidarity are likely to go up courtesy of minimization of opportunity for discord and competition. Teachers are only out to do their work and are not at loggerheads with anyone in the system.
- It minimizes discord and unnecessary competition among teachers which may bring the school into disrepute or chaos.

- It prevents display of superiority among teachers which can easily divide the teachers working in the same school.
- It covers other teachers' inferiority since they don't have to engage into a competition dog-fight with their colleagues.
- It keeps peace among teachers in school because they do what is within reach and do not mind other people's business.
- It promotes classroom autonomy since a teacher remains the alpha and omega in his/her class without the unnecessary evaluation.
- It builds self-confidence along the way as there is no opportunity for being put down by others. This comes as a result of keeping aloof from the possibilities of dissent and non-approval which take center-stage when people over interact.
- It leads to ideal self-evaluation instead of being evaluated by others. Since the best way of knowing yourself is by evaluating yourself, teacher insularity may help teachers to discover their strengths and weaknesses without having to refer to views of others.

Teacher insularity meanwhile possesses a *Negative Implication* on school systems in many ways. In a classic 1975 book, ''Schoolteacher'', Lortie described teacher isolation as ''one of the main structural impediments to improved instruction and student learning in public schools''. Lortie (1975) argued still that teachers have worked behind closed doors since at least the 19th century, and rarely if ever, collaborating with any of their school colleagues on improving on their teaching practice or examining students' work. Even the aspect of educational cohesion which is highly craved becomes a foregone conclusion (Farenga, 2013). Accordingly therefore, all can never be merry when teachers are not interacting at all good levels and the following stand out as key effects:

- Loss of constructive evaluation and criticism of their own accomplishments and professional support for their methods from their colleagues since they don't see one another at work and they often devalue the self-reports of each other.
- Development of self-doubt within teachers since professional confidence is largely self-constructed and therefore quite vulnerable, a few bad experiences or a couple of mediocre days from the teacher may create commotion. Even when things are moving well, you may never be sure about them yourself and you need approval of others.
- Failure to discuss common and crucial educational ideas based on aims and objectives and equally implied in the vision and mission of the school.

Fundamental questions will remain unanswered and there won't be a collective and collaborative handling of school wide problems.

- Absence of help even when desired by a teacher since the one giving the help will be skeptical that his suggestions and intentions may be misconstrued as drawing back. On a bad day, a teacher is likely to receive ridiculing advice to keep him off.
- The development of cliques among the teachers won't be avoidable in this case based on understanding since forging a collective understanding would have failed.
- Closely related to the above would be devising a method of survival within and this may be through composition of a language that is only understood by a few and used to shoot down others. Such jargons and slangs can be dangerous in a school setting.
- Failure to identify common learning problems among learners since every teacher will only be seeking to accomplish their own task(s).
- It kills off collaboration in a school system because no one is welcome to another's classroom or work station in the school. Teacher insularity in this case creates and feeds a stand-alone approach to work in the school system.
- Rumor mongering and backbiting become a permanent fixture in the school since, as already noted, cliques would have developed within teachers.

Complexities Encountered by School Managers / Leaders in Teacher Insularity-laden Situations

Since almost all teachers share feelings the same way, norms of interference are easily established in schools. Teachers in other words, are expected to stay out of another teacher's classroom. Such superficiality would only do little to promote sound educational policies, school programs or professional development and increases the isolation of the teachers whilst hindering development cohesion thereto. This creates a state of chaos in the school system which turns out to be complex especially from a leadership and management perspective.

Shyamala, P., contends that leadership during times of complexity and changes can be viewed from the perspective of various levels of complexity as presented below:

- Structural complexity relating to systems in the organization.
- Process complexity relating to the way things are done.
- Product Complexity relating to the nature of output in an organization.
- Managerial Complexity relating to the way people are led.

It is indeed the managerial/leadership complexity that this study focused on. How managers and leaders in the school system would fare in case of teacher insularity or isolation formed the basis for writing the paper in question. Management in the event of teacher insularity therefore, poses the following challenges or complexity:

- **Delegation of Authority:** This cannot take place in the wake of teacher isolation, attrition or insularity since the delegate will attract no audience whatsoever. Even when such an audience is available, respect from such would be hard to elicit. It should be noted however that delegation of authority is one of the most highly craved functions of management and without it, a leadership vacuum develops when a sitting leader retires, gets transferred, dies or better still has commitments.
- **Selection of a Leadership Style to Suit the School System:** This would be very hard on grounds that in cases of divided loyalty and identity, it is always unclear as to which leadership style suits best. The causes of insularity would also imply the failure of a one style fits all approach to leadership. The simple fact is that leading a divided people would be any leaders Achilles heel.
- **Creation of School Committees to Deal with Various Academic, Social, and Other School Programs:** This would definitely fall flat on the leaders. Committees require people who share a common understanding of various aspects that ascribe to school management and leadership but in the wake of teacher insularity and isolation, that would not be possible. Moreover without the committees, execution of school activities and programs would leave a lot to be desired.
- **Collaboration with Staff May Not Be Possible:** Given that every teacher will be attending to his/her own business, getting them together would prove futile. Without collaboration in the school system, ideas and philosophies that matter when it comes to school management would not be shared. This dents prospects of school tranquility and transformation.
- **Supervision of Staff Is Always a Key Challenge:** As most of the teachers in school will be doing things in isolation and secrecy, the leader would find it hard to supervise the staff. Moreover, knowing the true identity and feelings of the teachers would also prove futile since no one would be willing to open up to anyone. Improvement would thus be dealt a long blow indeed.
- Planning of staff programs may also become a one-man undertaking since having teachers on board to discuss a common aspect is always impossible. Planning for a divided house is planning to fail. With teacher attrition, isolation or insularity, everyone will be seemingly making their own plans leaving the school at the mercy of circumstances indeed.

- **Sharing of Information with Staff Becomes Complicated:** Since effective communication draws on information from multiple sources, circumstances involving teacher insularity would imply lack of communication flow. Moreover school activities and programs normally dwell on information and feedback. Without such, schools are bound to fail indeed.
- Collective decision making remains elusive in the school system since teachers would be drifting apart in all school undertakings. Decision making is premised on the assumption that different ideas come from different people and the most viable is selected for piloting and later, implementation. Within a teacher insularity dispensation, however, this would prove futile since each individual would keep cards to their chests.
- Motivation of staff may also be hard in such cases since it may be hard to establish the good performers in the midst of no approvals or evaluations of others. Staff performance evaluation is always possible through seeking views of others and then making ratings of the same. With isolation in full force however, teachers would not be in good position to do this thus making the work of the leader more daunting.
- Conduct of get together programs proves very complex since teachers in the school would not feel comfortable around each other and instead prefer virtually being alone. At the end of certain sessions in schools like school term, academic success based on release of examination results etc, schools seek to hold get together or celebration parties which ignite the spirit of togetherness and cooperation. With insularity and isolation in the mix however, this becomes elusive and leaders lose the opportunity to preside over staff based arrangements in the school.
- Reporting may become a complete spoiler of school systems: one of the ingredients of teacher insularity is actually enmity between teachers. Having said that therefore, it would be suicidal for one teacher to make a report on the other as the report would definitely be marred by negativity. The implication here would be a leader having to do everything on his or her own. Most crucially, even reports on students' progress may not be available and if they are, then they won't be that much trusted indeed given that the teacher making a report on a particular student may use the same platform to bring another teacher down.

Intervention Mechanisms to Teacher Insularity in Schools

Many authors have committed to exploring the various intervention mechanisms to teacher insularity (Heider, 2005; Hirsch, Koppich & Knapp, 2001; Maistry, 2008; Scholastic and the Gates Foundation; Steyn, 2013). Maistry (2008) actually opines that more studies need to be done to understand the conditions that would enhance and sustain collaborative structures for teachers in schools, especially in developing countries.

In response to this "high rate of attrition", have become a key tool for many state education associations the world over, seeking solutions to the problem of teacher isolation is indeed "formal mentoring programs". In fact, "formal mentoring" has not only become a very popular but also a wide spread teacher induction tool in recent years. In 2001, thirty-eight states were seen to be offering some kind of induction or mentoring program for new or novice teachers, as based on the view of Hirsch, Koppich & Knapp (2001).

Many approaches have always been earmarked to overcome the concept of teacher isolation or insularity as a mechanism of improving the quality of teachers and therefore the school system leading up to effectiveness of education:

1. The physical barriers can be eliminated by redesigning schools so that there is great potential for teacher interaction especially the classroom block design (Steyn, 2013).
2. The school administration should institute scheduled interaction and involvement of everyone in the department and even at school level. This can be done through collective scheming, meetings, curriculum review and development etc.
3. Assigning teachers leadership responsibilities intentionally tailor-made to suit consultation such that the idea deepens within them at all times. This will hopefully be extended to even the classroom teaching arena.
4. Mentoring programs in schools can also help teachers work with or deal with others in the system. That way, interaction becomes a *sine-qua-non* for all those in the system (Heider, 2005).
5. Peer coaching where individual teachers sharing common interests get to work together in the learning system. That way, and because all would share the same philosophy, teachers would feel free to interact.

More practical suggestions to combat teacher insularity are being proposed in this conceptual paper. The key ones thus reflect the following; arranging to eat lunch with a few colleagues for a regular schedule, creating a time where you merge your class with another teacher's (meet outside, in the cafeteria, or library to read together, do writer's workshop, or practice speech debate), if a secondary teacher, co-teach once in a while during your conference time (the other teacher can do the same, lending you a helping hand) and creating a walking club with a group of teachers and office staff. During lunch or recess, a walk around the block or campus may seem healthy indeed. On a positive note, teachers of the contemporary era are no longer wedded to the "cellular organization" as described by Lortie (1975). Indeed, based on the above, the potential value of collaboration for supporting school improvement through retaining skilled teachers would be recognized by all those charged with school management.

FUTURE TRENDS

Owing to the rates of teacher insularity prevalent in the education system today, employees and professionals ought to be moving towards the use of social tools for learning, working and teaching. Collaborating seamlessly face-to-face and at a distance, bringing the human element to virtual interactions, and personalized learning.

Even more of a driving force for teamwork and creating knowledge though, is our current culture which embraces a global mindset. To realize global ideas, there is need for collaboration and teamwork among all stakeholders. Understanding the power of collaboration holds the key to effective future systems especially in educational institutions.

Students will most likely drive the collaboration movement forward through peer projects, virtual study groups, and self-directed learning via their personal networks, though educators shouldn't be far behind. Research in such directions ought to be conducted ahead of such challenges if educational management and school leadership is to be effective.

CONCLUSION

Since almost all teachers share feelings the same way, norms of interference are easily established in schools. Teachers in other words, are expected to stay out of another teacher's classroom. Such superficiality would only do little to promote sound educational policies, school programs or professional development and increases the isolation of the teachers whilst hindering development cohesion thereto.

A number of education reform proponents have put unfailing great faith in the idea that ''teacher competition'' (e.g., merit pay) can render a dramatic improvement to educational outcomes. Nevertheless, despite such clear views and justifications, it is still early days to guarantee the effectiveness of such reforms, it is indeed greatly feared by many, an educationist that such policies will undermine ''teachers' collaborative work''. Ironically, competitive teacher assessment schemes that seek the best and worst in the school system premised on a rigid evaluation or rating system, could reinforce teacher attrition or isolation, so to speak. If teachers are competing with one another for merit pay, they cannot collaborate. They most definitely will go back behind closed doors and do work on their own without others getting the slightest of opportunities to peruse through. It is hoped that this is not the case because, as known, greater collaboration among teachers is a ''promising and important change in school structure'' and obtains positively in the realm of management and leadership.

This paper has argued and shown that teacher insularity, isolation or attrition cripples down the school system by affecting the major functions of management and administration on which the entire success and failure of schools rests. School leaders won't be in position to delegate, communicate, plan, motivate, share ideas, take collective decisions, organize get together programs etc. since teachers will not be willing to collaborate or cooperate with one another. The end result would be failure of the school and this would only be, but a delayed inevitability.

NOTE

Like in many other studies, in this analysis, equally, a number of terms have been used to bring out the concept of teacher insularity. The author is however acutely aware of the diverse meanings and interpretations to which the same concepts or terms can be subjected. The author thus moves in swiftly to insulate the study from those impending diversities by reducing the meaning of the same terms to an 'operational' level. The author thus claims that the terms shall only be referred to in terms of the definition offered here as far as this study is concerned. Some key terms underpinning the chapter are operationally defined in the "Key Terms and Definitions" section at the end of the chapter.

ACKNOWLEDGMENT

Am indebted to Prof. Dr. Şefika Şule Erçetin of Hacettepe University, Turkey for the invaluable and priceless inspiration, guidance and help throughout my course and career pursuit.

REFERENCES

Amin, E. M. (2005). *Social science research: Conception, methodology and analysis.* Kampala: Makerere University Press.

Ashton, P. T., & Rodman, B. W. (1986). *Making a difference: Teachers' sense of self-efficacy and student achievement* (p. 45). New York: Longman publishers.

Bisaso, S. M., & Rafiyyah, S. A. (2010). *The Relationship between perception of learning and teaching practices of lecturers: A study of educators at IIUM*. Malaysian Research Repository.

Cahill, K. (2010). *Even in chaos: Education in times of emergency*. Fordham University press.

Cunningham, R. (2000). *Chaos, complexity and the study of education communities*. Institute of Education.

Demir, M. K., & Ari, E. (2013). Öğretmen sorunlar- Çanakkale ili örneği. *Ondokuz Mayıs University Education Faculty Magazine*, *32*(1), 10.

Farenga, P. (2013). *Insularity of school research*. Retrieved from http://www.johnholtgws.com/pat-farengas-blog/2013/12/17/the-insularity-of-school-research

Glickman, C. D. (1985). *Supervision of instruction: A developmental approach*. Boston: Allyn and Bacon.

Heider, K. L. (2005). Teacher isolation: How mentoring programs can help. *Current Issues in Education, 8*(14). Available: http://cie.ed.asu.edu/volume8/number14/

Kara, A. (2008). *Understanding today's schools with Chaos Theory*. Academic Press.

Kasloff, P. (2007). *From teacher ısolation to professional collaboration: an experiment in school change*. Academic Press.

Lortie, D. C. (1975). *School teacher*. Chicago: University of Chicago Press.

McNeil, J. D. (1996). *Curriculum: A comprehensive introduction*. New York, NY: Harper-Collins Publishers.

McPherson, (1970). *Small town teacher. In sociological work*. Chicago: Aldine Publishing Company.

Mirel, J. & Goldin, S. (2012). *Alone in the classroom: Why teachers are too isolated*. Academic Press.

Murray, J. (2007). Countering insularity in teacher education: academic work on pre-service courses in nursing, social work and teacher education. *Journal of Education for Teaching: International Research and Pedagogy, 33*, 3.

Radford, M. (2006). Researching classrooms: Complexity and chaos. *British Educational Research Journal*, *32*(2), 177–190. doi:10.1080/01411920600568950

Shyamala, P. (n.d.). Leadership during times of complexity and changes: Human resource area. *Journal of Business and Management.*

Steyn, G. M. (2013). Building professional learning communities to enhance continuing professional development in South African Schools. *Anthropologist, 15*(3), 277–289.

Trygestad, J. (1997). *Chaos in the classroom: An application of chaos theory*. Roundtable Session of Chaos and Complexity Theory. American Educational Research Association. University of Minnesota. Retrieved from www.eric.ed.gov

Wasley, P. (2012). *The detriment of teacher ısolation.* Academic Press.

KEY TERMS AND DEFINITIONS

Chaos: Wikipedia defines the term chaos as any state of confusion or disorder while at dictionary.com it is defined as a state of utter confusion or disorder; a total lack of organization or order.

Complexity: The state of being complex or difficult with a variety of challenges and obstacles to effectiveness and efficiency. Complex situations are born of lack of a clear approach to dealing with an existing situation.

Insularity: The free online dictionary defines insularity as the state of being isolated or detached but the oxford dictionary looks at the term as ignorance of or lack of interest in cultures, ideas, or peoples outside one's own experience.

Leadership: The Oxford Advanced learner's dictionary of contemporary English opines that Leadership can be described as a process of motivating, preventing problems, ascertaining policies, principles and objectives, training or teaching subordinates, showing the way, learning how work shall be done, believing in team spirit and adopting empathy through love.

Unity: Wikipedia defines unity as the state of being undivided or unbroken. Meanwhile the free online dictionary considers unity as the state or quality of being in accord; harmony.

Chapter 10
Reflection of Chaos and Complexity Theory Properties within Classroom Discipline

Fathimath Saeed
Ministry of Education, Maldives & Gazi University, Turkey

ABSTRACT

Chaos and complexity theory has been used in the study of the natural sciences for over thirty years. Throughout the years, experts from various fields have used it as a new way to view the world around them, including its applications to the field of education and subsystems within the education system. However very few studies have been conducted on the application of chaos theory to classroom discipline. The field of classroom discipline, like the natural world, can also be observed from this perspective because it exhibits many features of chaotic/complex systems. Classrooms are often described as complex, dynamic and unpredictable environments. This makes it difficult for teachers to understand and manage classroom discipline. This paper explores the complexity of classroom discipline and how the principles of chaos and complexity theory reflect on classroom discipline. This would help inform management strategies and alternatives that would in turn enhance and improve student academic achievement and overall performance.

DOI: 10.4018/978-1-5225-0460-3.ch010

INTRODUCTION

The Complex Nature of Todays Classrooms

Today's schools are becoming increasingly diverse. As describes by *Tomlinson* and Allen (2000), today's classrooms consists of students from multiple cultures, some of whom are trying to bridge the languages and behaviors of two worlds. Students with very advanced learning skills sit next to students who struggle mightily with one or more school subjects. Children with vast reservoirs of background experience share space with peers whose world is constrained by the few blocks of their neighborhood.

As suggested by Howell & Nelson (1999), today's educators are expected to cater for the diverse needs of all students, including those with emotional and behavioural challenges (disorder). Some experts have referred this as inclusion. As defined by *Tomlinson* and Allen (2000), Inclusive education is about how we develop and design our schools, classrooms, programs and activities so that all students learn and participate together. Furthermore, it is also important to note, that the physical placement of students with special needs (disabilities) in general education classroom is not and end of itself, but rather a means to an end. The movement towards the inclusion combines with recent mandates requiring all learners to meet established curricular guidelines makes it increasingly challenging for educators to meet their moral and ethical responsibilities (Jones, Jones & Vermitte, 2013).

Technology is everywhere, incorporated in almost every part of our lives. It affects how we shop, socialize, connect, play, and most importantly learn. With their great and increasing presence in our lives it only makes sense to have mobile technology in the classroom. The use of technology in the classroom has become the most debating topic among the educators and decision makers (Edutopia, 2008). Traditional black boards have been replaced by smart boards, multimedia projectors. Students carry personal laptops and tablets instead of books. It is important that school wireless networks keep up with the ever changing technology in order to keep up with the students. From the ease of communicating with their teachers via e-mail, to quickly accessing an overabundance of information online about a particular topic they have learned about in class, technology is needed in today's classroom (Wainwright, 2015)

Discipline and Effective Classroom Environment

Discipline is one of the most important and useful skills everyone should possess. This skill is essential in every area of life, and though most people acknowledge its importance, very few do something to strengthen it. The success of teaching and learning depends on efficient management of classroom discipline. Well-managed classrooms provide an environment in which teaching and learning can flourish. As defined by Bear (2011), at community level, discipline could be the rules agreed by the members for the emergence of appropriate behaviors to ensure the wellbeing of their life in the community.

While Mahmud (2014), suggests, at the school level, discipline is the rules imposed on the students in order to control their behavior through punishment or reward. According to Slavin (2009), discipline includes methods to prevent or respond to behavior problems so they do not occur in the future. Discipline means, willingness to do something with the order, respecting the rights of others, practice good behavior without disrupting the interests of others. Discipline is the first step in creating and managing a positive learning environment. Discipline is an important component of the education system that is closely related to the ethics, morals and harmony (Slavin, 2009)

Classroom environment has a great impact on students' performance and achievement. Rutter & Maughan, (2002) argues that the learning atmosphere in the classroom can have a negative impact on pupil achievement. Researches have shown the working atmosphere in the classroom, and standards of pupil behaviour are major influences on how well pupils are likely to achieve. Van Tartwijk and Hammerness claims that learning is difficult, or impossible in a disorderly learning environment. The Hay McBer Report (2000) on effective teaching suggested that disruption and classroom climate were two of the most significant influences on pupils' learning opportunities and progress. The Hay McBer (2000) claims that effective teaching, and classroom climate are two of the most significant factors influencing pupils' learning opportunities and progress. Elliott and Phuong-Mai (2008) argue that several major studies on comparative international performance in education suggest that poor levels of classroom behavior may well account in part for the superior performance.

Undoubtedly, a positive learning environment plays a key role in pupil's achievement. The teacher is the agent responsible for creating the environment conducive for pupils learning. There are number of factors contribute to a positive learning environment, According to Wilen et al., (2004), part of building a supportive climate for learning involves teachers sharing their expectations concerning learning of content, achievement, and social behavior with their students.

As proposed by Dennis creating a positive learning environment involves the following as listed below:

1. **Building a Learning Community that Respects Individual Differences:** As mentioned earlier todays classrooms are filled with students from different social, economic and educational backgrounds and with different individual

needs. The teacher must acknowledge the differences among the individuals and teacher must establish a classroom environment that promotes understanding and tolerance for these different needs. As suggested by Strong (2002) one way to establish such environment is creating an atmosphere of mutual respect. Where students feel free to, clarify doubts; express their views and where tolerance and sense of common identity are promoted.

2. **Maintain Consistent Standards of Discipline:** As proposed by Shrunk, effective teachers establish strong classroom rules and follow through routines with the engagement of active and meaningful learning to minimize misbehavior and discipline problems.
3. **Create a Physically Appealing and Intellectually Stimulating Environment:** As suggested by Dennis, teacher should alter the physical setting or the arrangement of the class. By rotating the students and changing the rows and columns on a regular basis.
4. **Create an Environment that Promotes Silent Reading during Class:** This could be an effective activity that could promote learning.
5. **Create a Learning Community in My Classroom:** Dennis's learning community consist of students of same academic level and they work independently. The assignments targeted to these groups requires group discussion and later the writings are subjected to peer review.
6. **Promote Dispositional Thinking:** Dennis define dispositional thinking as intellectual qualities people possess that promote or cause intellectual development. Ritchart (2002) has grouped these dispositions into three categories: creative, reflective and critical thinking.
7. **Effectively Allocate Resources (Including Time):** Teacher should ensure that resources including material and time both are used efficiently and effectively.

CHAOS AND COMPLEXITY THEORY

Chaos theory came to prominence during 1980s. Edward Lorenz, a meteorologist was the first to talk about a known instance of chaotic behavior. He used a primitive model of global weather system that gave him a rough idea of how real weather behaved. He found out that very small changes in the initial conditions produced widely varying and unpredictable responses (Sardar & Abrams, 1999).

Complexity theory emerged from chaos theory and represents the body of research focusing on these systems that have complex characteristics. As suggested by Fahim and Talabari (2014), complexity theory deals with systems that display complex global behavior as a result of the local interaction of components, where the behavior of the components is determined by relatively simple rules and principles.

They further explain, like chaotic systems, the outcomes of these local interactions may not be linearly influenced by the initial conditions of the system, and so the prediction of the global action of a complex system cannot necessarily be dependent upon an understanding of the behavior of the lower-level components alone.

Chaos and Complexity Theory and the Classroom Environment

Complexity theory is also a theory of classifying problems based on how difficult they are to solve. Gleick (1987) and Morrison (1998) outlined the key principles of complexity: these principles are being examined in view of the classroom environment as follows.

- *Butterfly effect.* Small-scale changes in initial conditions can produce massive and unpredictable outcome (e.g. a butterfly's wing beat in the Brazil can produce a hurricane in America). An event in one school classroom can trigger off major incidents in the other school classrooms within a short time.
- *Very similar conditions may lead to very dissimilar outcomes.* A very general example of this could be the normal setting of the classroom. The teacher explains the lesson to the whole class but the level of achievement is different. They all cannot produce the same result.
- *Regularity and conformity break down to irregularity and diversity.* this could be explained in terms of the relationship between students and teacher and also between the fellow student groups. Somehow, the regularity and conformity breaks down and it leads to irregularity and diversity. Effects are not straightforward continuous functions of causes;
- *The universe is largely unpredictable.* the nature of the classroom environment is highly unpredictable. Classrooms consist of students from different family back grounds, culture and various ethnic groups. Students with different learning abilities including learning disabilities and physical disabilities could also be found in the same classroom.
- *If something works once there is no guarantee that it will work in the same way a second time.* the decisions made and actions taken on a similar situation cannot be at all or may not even work on another situation.

Doll (1993) also points out changes in the education system in the event of chaos and complexity theory which have a clear and direct attachment to the classroom environment. These critical aspects are the following:

- Determinism is replaced by indeterminism; deterministic, linear and stable systems are replaced by 'dynamical', changing, evolving systems and non-linear explanations of phenomena;
- Continuity is replaced by discontinuity, turbulence and irreversible transformation;
- Grand, universal, all-encompassing theories and large-scale explanations provide inadequate accounts of localized and specific phenomena;
- Long-term prediction is impossible;
- Order is not fixed.

Equally, social behaviour, education and learning are emergent, and are marked by recursion, feedback, evolution, autocatalysis, openness, connectedness and self-organization. As seen above, the classroom environment can be explained via complexity theory, because somehow our classrooms house the features of a complex environment. Maintaining the harmony and order of such a complex environment would be undoubtedly hard to achieve. The harmony and order can only be sustained by managing the classroom discipline.

Chaos and Complexity Theory and the Tenets of Classroom Discipline

Morrison (2002) states that the society is changing and the paradigms of understanding society are also changing. As schools and classrooms are sub social systems inside the bigger society, there is no way that classrooms can avoid these changes the society is undergoing. As suggested by Cohen et al (2011), change is ubiquitous and stability and uncertainty are non-concepts in complexity theory. Classroom environment, processes, interactions and strategies could be viewed through the lens of complexity theory (Ambrose et el 2014).

Equally, Ambros, Pierce and Sriraman (2014) suggest that principles of complexity such as internal diversity, redundancy, unpredictability and dissimilar outcomes from similar reaction could also be applied to today's classrooms to describe the complexity nature of the environment especially discipline.

The key elements of complex theory as suggested by Cohen et el (2011) include; Complex adaptive system, emergence, uncertainty and unpredictability, non-linear systems, diversity, networks, organizational learning, communication, order without control, self-organization, self-organized criticality, relationships, feedback, open system, holism, dynamic systems, distributed control and connectedness. A system in complexity theory can be described as a collection of interacting parts which function together as one unit and causes new elements to forms, new structures, new phenomena and new rules of behavior to arise (Morrison, 2002).

Larsen-Freeman (1997) points to 12 characteristics of the chaos/complex theory as follows:

- **Dynamic:** Changing over time, process rather than state.
- **Complex:** Having many parts, constantly acting and interacting.
- **Nonlinear:** Effect is disproportionate to the cause.
- **Chaotic:** A deep, coherent structure within apparent randomness.
- **Unpredictable:** Cannot forecast future states.
- **Sensitive to İnitial Conditions:** A tiny change can have a vast effect.
- **Open:** Energy/information can flow in or out.
- **Self-Organizing:** A structure/pattern emerges as components interact.
- **Feedback Sensitive:** Feedback is incorporated into behavior.
- **Adaptive:** Optimizes itself according to environment.
- **Strange Attractor:** Global pattern but unpredictable details.
- **Fractal:** A pattern that repeats itself at different scales (e.g. a tree).

Finch (2004) also explains some key concepts of complexity and systems theory and their relevance to language learning such as openness of system, connectivity, non-reducibility, emergent behavior and unpredictability and regularity.

The key characteristics of chaos and complexity theory mentioned above and how they relate with classroom discipline will be discussed as follows.

- **Non Linearity in Classroom Discipline:** As proposed by Larsen Freeman, 1997), nonlinearity means that the output of the system is not proportional to the input of the system and it may consist of synergistic reactions in which the whole is not equal to the sum of its parts. Fahim and Talabari 2014 describes, linear system as the sum of causes generates a corresponding sum of effects and that is enough to add the behavior of each component to make predictions and judgments about the behavior of the whole system. In such cases, small modifications lead to small effects, while great modifications and changes lead to large effects. In a linear system, a cause of specific strength results in an effect of equal strength.

Studies have shown understanding and dealing with classroom discipline are much more intricate and unpredictable than a linear setting would allow. Furthermore it's hard to predict and judge the behavior of whole class based on the single student and hence the behavior of single student cannot be generalized to whole class.

- **Sensitivity to Initial Conditions:** According to Salvatore, Indic, Murray and Baldessarini (2012), sensitivity to initial condition gave birth to chaos theory. It's a phenomenon discovered by Poincaré in his study of the *n-body problem,* a century later he proposed that randomness and determinism become somewhat compatible due to long-term unpredictability. Furthermore, a small difference in initial conditions may lead to huge difference in the final phenomena. Likewise a small error in the former may lead to an enormous error in the final stage. Therefore prediction becomes impossible and we see a random phenomenon.

Classroom discipline reflects features of sensitivity to initial conditions. A small and seemingly unimportant initial condition in dealing with discipline may create a bigger and unexpected outcome. Informing the teachers and students about such an effect may also help them in being more careful in dealing with discipline and misbehavior. Awareness of such an effect may also help students to think of their own strategies in dealing with the emotional effect of the initial conditions.

- **Dynamic Nature of Classroom Discipline:** It is believed that changing over time is an integral property of a system. One of the major principles of chaos theory is that complexity theory is the science of process rather than a state (Gleick 1987). Cambridge dictionaries online defines dynamic as continuously changing or developing.

Classroom discipline can be viewed as a system with numerous subsystems attached to it which is subjected to continuous variation at any given time.as a result the main system is also subjected to continuous to change. For example the sub systems could be the environment, teaching and learning strategies the physical and biological features of the students and cognitive strata they own. This may lead the discipline to be a process rather than a state.

- **Complexity and Classroom Discipline:** As suggested by Fahim and Talabari (2014) chaos system s are complex due to 2 major reasons. Firstly, complex systems often contains large number of components of agents. Secondly, the behavior of complex system is more than a product of the behavior of its individual constituents. That is the behavior of complex system emerges from interaction of its agents with each other. Each agent is constantly acting and reacting to what the other agents are doing.

Discipline is complex as it contains large number of agents that are interdependent and these agents are in continuous status of interaction with each other and constantly reacting and acting to what other agents are doing. Students act and react to what society portrays, what the school community is doing etc.

- **Connectivity and Classroom Discipline:** Larsen-Freeman (1997) asserts that "Complete Interconnectedness" is one of the characteristics of a complex system. All variables are interrelated, and therefore changes in one variable will have an impact on all other variables that are part of the system. Consequently, complex systems cannot exist in isolation, but affect each other in a multitude of ways.

In terms of the classroom discipline, everything occurring in the classroom can have impacts on everything else. The teacher's use of punishment and reward as to modify behavior is an example of this concept.

- **Unpredictability and Regularity of Classroom Discipline:** A major reason for the unpredictable behavior of complex systems is their sensitivity to the initial conditions. Because we can know all the initial conditions of a complex system in details, so we cannot predict the ultimate outcomes of a complex system. Even very small errors in measuring the state of a system will be augmented dramatically, making any prediction useless (Sardar & Abrams, 1999; Retrieved from http://www.Whatis Chaos Theory FractalFoundation.org.html).

If we apply this to classroom atmosphere especially in the context of discipline the same trend can be observed. Cvetek (2005) discusses that every teacher knows that in spite of the best prepared lesson plan, it is impossible to predict what will happen at any given time in the lesson and thus prepare in advance what may be the best teaching decisions in that lesson. No matter how prepared the teacher is to tackle discipline in the classroom it's difficult to predict it. Because, the initial conditions are unknown thus, making it impossible to predict the behavior and ultimate outcome.

- **Determinism of Classroom Discipline:** Chaotic systems are deterministic, but not predictable. Their unpredictability does not mean that they are random or indeterministic. Determinism is a belief that assumes every outcome is a result of preceding action (Fahim and Talabari, 2014). Sardar & Abrams, 1999 agrees that due to sensitivity to initial conditions, the behaviour of chaotic and complex systems is unpredictable even though it is deterministic.

Since, classroom discipline is out comes that results from preceding actions of students or teachers. Could be actions that occur while teaching, lesson planning, learning and etc.

- **Self-Organization and Adaptive Nature of Classroom Discipline:** Self-organization denotes a process by which a structure or pattern emerges in an open system without specifications from the outside environment. When a system of this type receives a sufficient amount of energy, it may become unstable. As a result of this instability, an originally uniform state can give rise to a variety of complicated temporal, spatial and behavioural modifications. Larsen Freeman (1997) proposes the capacity to self- organizes shows that complex systems are adaptive to the changes and modifications these systems acquire patterns of behaviour without any input from the outside world. Adaptation is evolution. To survive in the real world, one must continually adapt to the challenges that life presents.

Classroom discipline is self-organizational, open and an adaptive system. Self-organizational as the nature and types of discipline problem varies without any influence or pressure from outside world. However the existing problems do adapt to fit in to the present system.

- **Emergence Nature of Classroom Discipline:** Many complex systems display the feature of emergence. According to Fahim and Talabari (2014) emergence is the patterns of behaviour that arise as a result of simple interaction between the individual agents of complex system. These systems, which are basically chaotic, or complex, have the capacity to generate patterns that are seemingly non-chaotic or predictable in behavior. Furthermore, they explain this feature of chaotic systems using weather as an example. It is difficult to identify the initial conditions that trigger individual weather patterns or to predict the specific outcome. However the emergent patterns such as cloud formation can be used to predict the overall behaviour of the weather system.

Considered from a classroom discipline perspective, one realizes that students' behavior emerges naturally despite the set of rules and regulations put across. Even when teachers try to control some behavioral aspects of learners, they still happen in the school system.

- **Feedback Sensitivity within Classroom Discipline:** As described by Sardar & Abrams, 1999 it is a specific property of any system where output or the result alters the input of the system. It serves as a messenger from the outside, which informs the system of what it needs to do to continue existing. The systems that modify their behavior according to feedback continue to evolve, adapt, and thrive; otherwise, they will likely stagnate and disappear.

Classroom discipline is susceptible to feedback, teachers and school leaders. Without feedback, the stakeholders in the school system may not be in position to manage the learners and their discipline to an efficient and effective level.

CONCLUSION

Taking the above analysis into consideration, classroom discipline exhibits complex characteristics. It's a complex global behavior as a result of interaction of components. Discipline is a behavior due to interaction of teachers, students, learning and teaching. Chaos/complexity theory has the potential to contribute to our awareness about various aspects classroom discipline.

The deterministic view of education that classrooms and discipline are simplistic, cause-effect systems which can be easily manipulated, quantized and controlled is not addressing the problems of today's schools. From an alternative perspective, chaos and complexity theory gives us an understanding that the things we consider unimportant or trivial in our daily lives might have an equal weight in terms of affecting the results as the things we consider important.

Teachers ought to be abreast with knowledge of chaos and complexity in order to be able to deal with classroom discipline effectively. The theory would impress it upon teachers that where seems to be chaos equally lies a key aspect of orderliness that can be adopted to transform the system.

REFERENCES

Bear, G. (2011). *Discipline: Effective school practices. National Association of School Psychologist*. Newark, NJ: University of Delaware.

Canter, L., & Canter, M. (1976). *Assertive discipline: A take-charge approach for today's educator*. Los Angeles, CA: Canter and Associates.

Charles, C. M. (2002). *Building classroom discipline* (6th ed.). New York: Longman.

Coloroso, B. (2002). *Kids are worth it! Giving your child the gift of inner discipline*. New York: Harper Collins.

De Bot, K., Verspoor, M., & Lowie, W. (2005). Dynamic systems theory and applied linguistics. *International Journal of Applied Linguistics*, *15*(1), 116–118. doi:10.1111/j.1473-4192.2005.0083b.x

Edutopia. (2008). *Why Intergrate Technilogy into the curriculum? The Reasons Are Many*. Available on www.edutopia.org

Fhim, M. & Talabari, A.F. (2014). Chaos and complexity theory and education. *The Journal of English Language and Learning*, 13.

Glasser, W. (1965). *Reality therapy: A new approach to psychiatry*. New York: Harper & Row.

Glasser, W. (1977). 10 steps to good discipline. *Today's Education*, *66*, 61–63.

Glasser, W. (1986). *Control therapy in the classroom*. New York: Harper & Row.

Glasser, W. (2009). *Classroom Management: Dealing with Discipline Problems*. Quality Education Programs, Inc. San Pedro.

Gleick, J. (1987). *Chaos: Making a new science*. New York: Penguin Books.

Jones, F. (1979). The gentle art of classroom discipline. *National Elementary Principal*, *58*, 26–32.

Kohn, A. (1996). *Beyond discipline: From compliance to community*. Alexandria, VA: Association for Supervision and Curriculum Development.

Kounin, J. S. (1970). *Discipline and group management in classrooms*. New York: Holt, Rinehart & Winston.

Larsen-Freeman, D. (1997). Chaos/complexityscience and second language acquisition. *Applied Linguistics*, *18*(2), 141–165. doi:10.1093/applin/18.2.141

Morrison, K. (2006). *Complexity Theory and education*. APERA Conference 2006, Hong Kong.

Morrison, K. (2008). Educational Philosophy and the Challenge of Complexity Theory. *Educational Philosophy and Theory*, *40*(1), 19–34. doi:10.1111/j.1469-5812.2007.00394.x

Salvatore, P., Indic, P., Murray, G., & Baldessarini, R. J. (2012). Article. *Dialogues in Clinical Neuroscience*, *14*(4), 369–379. PMID:23393414

Saphier, J., & Glower, R. (1997). *The skillful teacher: Building your teaching skills*. Research for Better Teaching, Inc.

Sardar, Z., & Abrams, I. (1999). *Introducing Chaos*. Icon Books Ltd.

Seyyedrezae, H. S. (2014). The application of chaos/complexity theory in classroom teaching, task design and lesson planning. *Journal of Language Sciences & Linguistics*, *2*(2), 27–32.

Skinner, B. F. (1968). *The technology of teaching*. New York: Appleton-Century-Crofts.

Skinner, B. F. (1971). *Beyond freedom and dignity*. New York: Knopf.

ADDITIONAL READING

Baker, P. H. (2002). The role of self-efficacy in teacher readiness for differentiating discipline in classroom settings/ Doctoral dissertation, Bowling Green State University. *Dissertation Abstracts International*, *63*, 1783.

Bandura, A. (1993). Perceived self-efficacy in cognitive development and functioning. *Educational Psychologist*, *28*(2), 117–148. doi:10.1207/s15326985ep2802_3

Bradley, R., Henderson, K., & Monfore, D. A. (2004). A national perspective on children with emotional disorders. *Behavioral Disorders*, *29*(3), 211–223.

Brouwers, A., & Tomic, W. (1999). The factorial validity of the teacher interpersonal selfefficacy scale (Report SP038950). The Netherlands: Netherlands Open University. (ERIC Document Reproduction Service No. ED437365)

Bullock, L. M., Ellis, L. L., & Wilson, M. J. (1994). Knowledge/skills needed by teachers who work with students with severe emotional/behavioral disorders. *A revisitation. Behavioral Disorders*, *19*(2), 108–125.

Curwin, L. R., Mendler, A. N., & Mendler, B. D. (2008). Discipline with dignity: New challenges and solution. (3rd Edition) Association for Supervision and Curriculum Development Friend, M., & Pope, K. L. (2005). Creating schools in which all students can succeed. Kappa Delta Pi Record, 41(2), 56-61.

Curwin, R. L., & Mendler, A. N. (1997). *As tough as necessary: Countering violence, aggression, and hostility in our schools*. Alexandria, VA: Association for Supervision and Curriculum Development.

Fuchs, L. S., Fuchs, D., & Bishop, N. (1992). Instructional adaptation for students at risk. *The Journal of Educational Research*, *86*(2), 70–84. doi:10.1080/00220671.1992.9941143

Gibson, S., & Dembo, M. H. (1984). Teacher efficacy: A construct validation. *Journal of Educational Psychology*, *76*(4), 569–582. doi:10.1037/0022-0663.76.4.569

Hamill, L. B., & Dever, R. B. (1998). Preparing for inclusion: Secondary teachers describe their professional experiences. *American Secondary Education*, *27*(1), 18–27.

Hersey, P., Blanchard, K. H., & Johnson, D. E. (1996). *Management of organizational behavior: Utilizing human resources* (7th ed.). Upper Saddle River, NJ: Prentice Hall.

KEY TERMS AND DEFINITIONS

Butterfly Effect: Small-scale changes in initial conditions can produce massive and unpredictable outcome.

Chaos: Complete disorder or confusion.

Classroom Discipline: The appropriate action by a teacher toward a student or group of students, after the student's behavior disrupts the ongoing educational activity or breaks a pre-established rule created by the teacher, the school administration or the general society.

Classroom Environment: The intellectual, social, emotional and the physical environment in which students learn.

Classroom Management: How a teacher organizes his or her students, materials, time, space, teaching and learning, so that the students can learn in a proper environment.

Complexity: Used to characterize something with many parts where those parts interact with each other in multiple ways.

Conformity: A social influence involving a change in belief or behavior in order to fit in with a group.

Determinism: Philosophical idea that every event or state of affairs, including every human decision and action, is the inevitable and necessary consequence of antecedent states of affairs.

Effective Classroom Environment: The environment in which teaching and learning could flourish.

Indeterminism: The concept that events are not caused, by prior events.

Nonlinearity: Output of the system is not proportional to the input of the system.

Chapter 11
Corruption as a Challenge to Quality and Quantity of Education in Sokoto State, Nigeria

Mukhtar Salihu Nawait
Sokoto State University, Nigeria

ABSTRACT

Corruption has caused untold havoc not only to the quality, quantity of education and credibility of citizens in Nigeria, but also to our national growth and development in all aspects of human endeavor. In academia it diffuses and circulates through the aorta and capillaries of the system, such that it is no longer view as a bottleneck thwarting the successful implementation of educational policies. The ugly scenario has degraded our educational foundation popularly called lower basic created a permanent vacuum which the subsequent higher level proved helpless to fill. The objectivity, reliability, trust, transparency, sincerity and above all integrity known in the academia as its "glory" have been wiped out by corrupt practices. At this junction, it is mandatory to dig out the reasons behind indulgence in corruption in our lower basic, identify its causes, approaches and effects as they are dimensional in nature. The paper will attempt the definitions of the term corruption at initial, and the concluding aspect will suggest the possible remedies.

DOI: 10.4018/978-1-5225-0460-3.ch011

INTRODUCTION

The oldest industry known to man is education. Indeed, education is the only indiscriminate vehicle for social mobilization and dynamic civilization, where poverty gives way to affluence, disease is eradicated and replaced with a healthy vibrant populace, ignorance's gives way to scientific, rational and progressive thinking which will collectively catapult the rapid growth of any society, not only in term of economic growth but also in political maturity. This is because the sociopolitical and economic development of any nation is determined by the quality and level of educational attainment of the population. Undoubtedly however, basic education is the bedrock of all tiers of subsequent stages of education which needs to be given all the attention it duly deserves, if we really want to have good products at both secondary and tertiary institutions.

"Nigeria has toiled with some educational programs, which have only served as conduits to transfer money to the corrupt political leaders and their cronies. For instance, the nation launched the *Universal Primary Education* (UPE) in 1976, but as noted, the program failed due to lack of fund necessitated by corruption, among other factors. Nigeria has again launched another mass-oriented education program, this time branding it the *Universal Basic Education (UBE)"*. The President, *Olusegun Obasanjo,* declared during the launching of the program in *Sokoto* that the nation "cannot afford to fail this time around." However, not long after that, the federal government reported that the falling standard of education in Nigeria is caused by "acute shortage of qualified teachers in the primary school level." It is reported that about 23 percent of the over 400,000 teachers employed in the nation's primary schools do not possess the necessary teaching qualifications, even when the National Certificate of Education (NCE) is the minimum educational requirement one should posses to teach in the nation's primary schools" (*Ogbeifum* & *Olisa*, 2001).

Unfortunately, none of the strategies so far adopted has been able to yield any desired result. The most disturbing aspect of it is that the people who are supposed to be actively involved in the crusade against this evil monster to the success of the program are instead aiding and abetting it in one way or the other. These are board officials, administrators, head teachers, parents as well as the general society.

The Concept of Corruption

According to Advanced Learner Dictionary "corruption refers to dishonest or illegal behavior especially of people in authority". In other words it refers to the changing of moral act to immoral. The Oxford Dictionary "defines corruption as (of person or their actions), immoral, dishonest, the offering and accepting of bribes, impair, debase by errors or alterations, decay". In a more social and political context, corruption "involves inducement intended to pressure an official from carrying out his functions in accordance with set rules and procedures. It includes act of arrogation of benefits for one self or ones friends and relatives to the detriment of the right or entitlement of others regarding the same or similar benefits" Yusif, (2001).

However it is generally accepted that corruption "involves inducement intended to pressure an official from carrying out his functions in accordance with the set rules and procedures". "It includes act of arrogation of benefits for one's self or ones friends and relatives to the detriment of the right or entitlement of others regarding the same or similar benefit, thus to be corrupt is to be opened to bribery and dishonesty and to be immoral" (Yusif, (2001).

Anechiarico and Jacob (1996) and Kalnin (2000) cited in (Nawait et. al, 2009) state that the definition of corruption derived from the more general corrupt issue, like other areas, it include the abuses of authority for material gain. In education since professional standard is equally important, hence the definition of corruption include the abuse of authority for personal as well as material gain.

The Universal Basic Education, UBE, came as a replacement of the Universal Primary Education and the primary is replaced with lower basic. It is an innovation designed to enhance the success of the first nine years of schooling. The UBE involves 6 years of Primary School education and 3 years of Junior Secondary School education, culminating in 9 years of uninterrupted schooling, and transition from one class to another is automatic but determined through continuous assessment. This scheme is monitored by the Universal Basic Education Commission, UBEC, and has made it "free", "compulsory" and a right of every child. Therefore, the UBEC law section 15 defines UBE as early childhood care and education. The law stipulates a 9-year formal schooling, adult literacy and non-formal education, skill acquisition programs and the education of special groups such as nomads and migrants, girl child and women, Al-majiri, street children and disabled people (Aderinoye, 2007), cited in UBEC, (2012).

Oxford Dictionary defined Quality education as "the development of character or mental power. To support this definition, Essien (1975), stated that, quality education is the education that enables the educated person to posses qualities of the mind and the knowledge of the purpose of life and the twin requirement of an educational experience which is the ability to use the developed skills and knowledge. It is the type of education that enabled the individual to interact effectively with the challenges within and outside his/her environment. The purpose of education is not rested only on the acquisition of skills and experience, rather their application to provide solution to societal problems like eradication of ignorance, reduces poverty and enhancement of social economic values and fulfillment of moral responsibilities.

While the concern for quantitative achievements is justified, the increasing neglect of quality aspects could be highly damaging. Quality looks like being threatened by focusing rather exclusively on quantity.

To provide sanity within the society, corruption needs to be completely wiped out. And such crusade most starts from the education sector. It is very necessary to clean the educational sector since education is the second or third most powerful and effective instrument of socialization. Indeed, after the family and possibly religion, education follows as the most important agent of socializing children and the youths generally. As a result, if children and the youths are not properly socialized, they are very likely to end up adopting destructive social values.

Approaches of Corruption at Lower Basic

Corruption and poor governance are acknowledged as major impediments to realizing the right to education and to reaching the Millennium Development Goal of universal basic education by 2015. Corruption not only distorts access to education, but affects the quality of education and the reliability of research findings, and manifestations of corrupt practices repeat themselves across Nigeria (Routledge, 2013).

These include;

- **Planning and School Management:** Decision on government funding for new and existing schools are often taken outside the appropriate organs, outcomes include unnecessary building of schools as project are selected for purely personal and political reasons, disregarding real needs. Corruption in procurement affects the acquisition of educational material curriculum development, textbooks, library stock, uniforms, etc.), meals, building and equipment. As sales levels are guaranteed in such acquisition, bidders eagerly pay bribes to secure the infallibly high profits. Where textbooks and supplies are monopolized by the state and bidding procedures are irregular poor quality provide become the norm and contracts are frequently secured by unprofessional agents. Moreover private schools that have not attained the standard may bribe their way in getting necessary authorization and corrupt accreditation.
- **Students Admission and Examination:** Admission of pupils into lower basic schools and entrance examination papers can be sold in advance to high-paying candidates. Oral examinations are even more open to corruption as evaluations are subjective and difficult to monitor. Low quality school; turn out student who are ill prepared for college, thus forcing parents to hire private tutors to ensure that their children pass the entrance exams. Supposedly

from lower basic, education becomes prohibitively expensive for poor families when the reality requires paying for private tutors in order to pass or admitting pupils in private schools which is more expensive. The above factor contribute to the student demand on their traditional "right" to cheat which arose the problem of beatings and killings of conscientious staff members who attempt to work honestly during the examination and those that put effort in stopping the sales of the exam questions.

- **Teachers Recruitment and Professional Conducts:** As criteria for recruitment are repeatedly bypassed, unqualified personnel are often appointed. Placement in rural schools tends to be popular, especially among unmarried and female teachers and can sometimes be avoided by bribing public officials. Skewed distribution of teachers posting can leave some schools overstaffed and others in crises. Also salaries paid to "ghost teachers" are a source of problem which worsens with a proliferation of rules and regulations.

Motivated and efficient teachers are crucial for quality in teaching. However, people in developing countries soften complain of absent or abusive teachers and their demands for illegal fees. It is not uncommon to find pupils exploited as unpaid labour, drunken teachers in school, sexual and physical abuses are common and in most instances teachings is conducted at all (Nawait, et. al 2009). The above stated approaches vividly discussed by (Routledge, 2013, are applicable in all part of Nigeria Sokoto state inclusive, but the causes of corruption are numerous and dimensional and therefore differ among individuals, groups or societies.

- **Causes of Corruption at Lower Basic:** Apart from various approaches of corruption, there are numerous causes working collectively and simultaneously in perpetuating corruption at lower education. These causes that are social, political, cultural as well as educational in natures generate a lot of negative impacts that ruin education system. However, "the *UNICEF* in its 'state of the world's children' report for 1999' pointed out that about four million Nigerian children have no access to basic education, and that majority of those that are 'lucky' to enter schools are given sub-standard education" (*Akhaine*, 1999, cited in UBEC, 2012).

Unpatriotic feeling toward infrastructural development, poor maintenance and rehabilitation culture and negligence attitude toward schools facilities are the common characteristics of most corrupt education officers. The outcome yielded out insufficient classes to accommodate students for learning. The existing classes in schools are over-crowded with few students on seats and majorities are either hang on open windows or sitting on bare floor. It's exceedingly difficult for students to concentrate on their studies in such unbearable condition. In schools where furniture are claimed to be provided, it is common to found five to six students sharing the same two-seater. These inhuman attitudes thwarted the quality and quantity of education at lower basic in the Sokoto State.

Lack of transparent regulation and criteria couple with embezzlement of public funds plays a significant role in stimulating and promoting of corruption at lower basic of education. Large amount of money are usually budgeted to shoulder many capital and recurrent expenditure in primary education, but greater percentage of that money may not be spend on the purpose that were budgeted for. For instance, in the year 2005 alone, Sokoto state government launched an operation to resuscitate primary education which consumed N6.5 billion naira for construction of 4500 additional classroom, (Junaidu 2005). But two years later, many of such classrooms were off-roof or even collapse. Just because corruption provided an enabling environment for the two parties: the authority and the contactors to negotiate on illegal beginning of N6.5 billion, (Nawait, et. al 2009).

Some cultural practices can aggravate the problem of corruption at lower basic of education. For instance, there are sometimes strong link between social corruption and corruption in academic. In this case, authority shares a lot of blames. In 80s and early 90s an educationist can only have a dream of becoming a secretary of education in local authority after attainment of certain experience despite his volume of professional qualifications. He has to be a classroom teacher, head teacher, supervisor, inspector and possibly monitoring officer. But now, a fresh graduate lacking adequate educational qualification and experiences may be given an automatic offer to take over the same office. How can this category of secretary cope with the challenges filed up in the local education authority? The achievement of such type of leader may not exceed improper recruitment and posting of teachers as well as drastic falling in educational standard and mass failure in national and regional examination which are the order of the day.

However, parents are also the agent of falling standard of their children education. Time without number, parents' form the habit of influencing vertical promotion of their children, for instance, an average primary 5 pupil push to JSS 1 may likely face serious academic problems. This attitude become a common practice in our education system. Affected child has to work extra or else he will slide down below average. In some cases parents dictate the career of their children. "I want my child to be a medical doctor or a pilot without considering his mental and academic capability". Virtually this contributes poor quality of basic education and subsequently the other levels.

Teachers are not left behind; corruption has circulated within their blood and engulfed all the elements of punctuality and deposited absenteeism. Absenteeism has become a common attitude of most teachers. These teachers indirectly are giving pupil's idea that one can have a job without working. In fact it creates permanent corrupt way of thinking in the fresh mind of innocent children. It is important to note that, the greater the number of hour's teacher learner interaction the more the child will benefit. For instance, Japan has an academic year of 286 days, US and most countries in Western Europe have academic years of 280 days at average, but poor developing countries like Nigeria have an average of only 220 days which may suffer for series of public holidays and necessary breaks (Iguzor, 2006) cited in (Nawait, et al., 2009). The result of the above problem most includes carryover of uncovered syllabus year in year out which in turn affects the totality of this system.

The poor attitude of most governments towards education and lack of proper management and utilization of public fund as a result of ineffective monitoring and supervision catapult the falling standard of basic education. This ugly situation gives private school enough breathing space.

The proliferation of private schools serves as an index signifying increase in falling standard of education of public school. Even with the establishment of UBE in 2004 by Act of National Assembly and its declaration of free and compulsory basic education, people are still trooping toward private school. This indicates that the impact of UBE is yet to be felt by population.

All these, including ubiquitous corruption, have contributed to the decline in the quality of instruction in Nigeria's educational institutions that were ones highly regarded. With the news of corruption still filling the pages of Nigeria newspapers and magazines, the apparent war on corruption in the society seem an impossible task, since those wagging the corruption-war are themselves as corrupt as a parrot. Poor schooling, ignorance, poverty, and unemployment or underemployment among the youths could lead to their being easily manipulated by the political elites for selfish purposes. That will spell danger for the society, as this group will become the nation's leaders of tomorrow (UBEC, 2012).

Effects of Corruption

In Sokoto state like other states in the northern part of the country, corruption gives policy makers room to plan and approved fake decisions for education, executives to award and implemented contracts of billions in paper, teachers and students with qualifications that are subjective and parents spending huge amount money for fake result for their siblings. This clearly indicated that, corruption has denied us access to all positive aspects of our life. "The greatest cost of corruption in the education sector is the denial of young people reaching their full potential and becoming positive, contributing citizens to their communities and society. Given that education is the major driver of personal and social development, corruption in education guarantees an uncertain future by denying people their dignity, rewarding achievement without merit, and normalizing societal mistrust. Sustainable futures for the global community are in peril if corruption breaks down the political, social, and moral structures of nations".

The Global Corruption Report: "Education released by Transparency International frames corruption in education within the context of global trends and explains why education is vulnerable to abuse of power for private gain. In most nations, the education sector receives the largest or second largest percentage of public funds and employs the greatest number of public servants. According to the report, in countries such as Ethiopia and Indonesia, public spending on education exceeds more than a quarter of total public expenditure. This is a strong indicator that education is valued for its transformative contribution toward human and national development, yet this intrinsic understanding and the amount of money involved for financing makes education a high-stakes endeavor. Corruption in education will prevail for many middle- and low-income countries, which do not have the national capacity to appropriately manage financing, strong national and local administrative systems and structures, stakeholders invested in quality education services, or accountability systems" (Alifa, 2001*)*.

Nigeria's school compounds today are like facilities located in war zones. They are in a serious state of disrepair, more comparable to makeshift refugee camps in a war zone with broken windows, chairs, and desks. In some schools, the roofs are leaky and when it rains, classes are interrupted. This horrible condition is what the corrupt leaders are interested in so that, more budgetary allocation will be possible (Priye 2011).

In the 1960s, 1970s, and 1980s, vice chancellors, registrars, deans, chairs, principals, headmasters and professors/lecturers/instructors/teachers closely identified with their students and encouraged them to perform academically and assist them in developing their social skills. In fact, school administrators and teachers/instructors/professors treated the students as their own children and were committed to ensuring that students completed their education with flying colors. Today, it appears that many school administrators and teachers/instructors/professors are more interested in exploiting the students for their own financial gains (Priye 2011).

It is obvious that the educational sector in Nigeria and Sokoto state in particular is currently full of black sheep breed, deployed and sponsored by corrupted officers for their selfish material gain. The sanity and faded glory in the sector will only restore if such undesirable elements are cleaned or formatted. "Therefore, the war on corruption cannot be won without making a determined effort to purge the educational sector of the psychosocial beast. It appears that there is a symbiotic relationship between the educational culture and the psychosocial and political culture of the society at large. What happens to society at large affects the educational sector and whatever happens in the educational sector affects the society at large. Those in the educational sector learn and adopt predominant values generated by society at large and the society at large learns and adopts predominant values generated by the educational sector. It is very necessary to clean the educational sector since education is the second or third most powerful and effective instrument of socialization. Indeed, after the family and possibly religion, education follows as the most important agent of socializing children and the youths generally" (Priye 2011). However the following future suggestions are put across to provide sanity in education sector.

CONCLUSION

Despite the World Bank Programme vision on Universal Basic Education (UBE) for all by 2015, with this presence level of corruption in education sector, such vision will remains a cock and bull story. The only alternative for the vision is to fight corruption and its related practice whole heartedly and honestly. To wiped out or minimize the menace of corruption in sokoto and Nigeria as a whole, parents, stakeholder, communities, government and non-governmental organisations should co operate earnestly to put a stop by contributing actively in fighting corruption. Therefore, if anti-corruption agencies are empowered, appointments, recruitments and deployments of teachers are on merit and students' admissions, placement and sponsorship are also on merits, corruption will reduce to its barest minimal .

FUTURE TRENDS

The anti-corruption agency offices with patriotic officers should be opened and empowered at all level of government, particularly education ministries and boards to deal securely with corrupt cases. They should also have legislative backing to operate independently and work directly with judiciary.

Teachers and non academic staff should be paid a reasonable amount that will enable them to live above poverty level to discourage corruption in the education sector. This can be achieved by readjusting the salary scale of teacher to be uniform regardless of tears of government.

Appointment, redeployment and transfer should be base on merit and objective's need, rather than selfish fulfillment of political motives. Teachers should have job security and job assurance at all level in order to be free from persecution, prosecution and manipulation in the hand of political holders.

Award of Scholarships from government and other donor agencies should targeted students from less privileged and poor family background in order to reduce the existing gab between children of have and have not. It is indeed a corruption in the highest order to sponsor children of the rich and denied those of the poor.

Government should renew its education loan agreement with commercial bank for students to enable poor parents and guardians secure reasonable fund to meet the financial need of their wards especially those studying professional courses like medicine and engineering.

Provide Jobs for Graduates of Educational Institutions: Government should provide jobs for outputs of educational institutions and grant loans with reduced interest to graduates that are willing to be self-employed, to curtail the rising trend in unemployment in the country.

Since teachers are the chief implementers' of all educational programme particularly the curricula, their recruitment should be base on merit at all levels. There is growing need of more and more qualified teachers to Cushing the effect of student boom in Nigerian's schools. There is also need to arrange for teachers welfare packages especially those deployed in rural areas to enable have comfort and satisfaction in teaching profession.

Teachers should also be encouraged to attend seminars, conferences, and workshop grants and sponsorship to update their knowledge and improve their skills.

REFERENCES

Alifa, D. (2001, June 17). Intrigues in FG-ASUU Face-off. *The Guardian.*

Dike, V. (2001). *Who is the author of democracy and political life in Nigeria.* Zaria: Ahmadu Bello University Press.

Essien, J. E. (1975). *Education and challenges of quality.* Benin Ethiope Publishing Corporation.

Nawait, M.S., Sirajo, A., & Nasir, Z.M. (2009). Multi-Disciplinary conference. *Farfaru Journal, 4.*

Ogbeifum, S. & Olisa,E. (2001, July 1). Half-baked teachers bane of education woes. *The Vanguard Online.*

Oladeji, B. (2001, August 2). More private varsities coming FG. *The Nigerian Tribune.*

Onuorah, M., & Aniete, B. A. (2001). Govt Releases N1b To Pay Police. *The Guardian.*

Priye, S. T. (2011). *The corrosive effect of corruption on Nigerian educational system*. Academic Press.

Reuters. (2001). *Nigerian says killed minister for one min naira*. Ibadan, Nigeria: Reuters.

Rousseau, J. J. (1994). *The Social Contract*. Oxford University Press.

Routledge. (2013). *Global corruption report: Education transparency international*. Author.

The Guardian Online. (2001, August 2). *Money and politics of bitterness*. Author.

UBEC. (2012). About UBEC. Universal Basic Education Commission. Retrieved 30 August 2012.

UBEC. (2012). *About UBEC*. Universal Basic Education Commission.

Umar, B. & Adoba I. (2001, December 6). Senate Passes UBE Bill. *ThisDay*.

UNESCO. (2011). *World data on Education*. UNESCO-IBE.

Vocational Education in Nigeria. (2012). UNESCO-UNEVOC.

Yusif, M. M. (2001). *Education and the war against corruption in Nigeria: some basic issues on the role of primary education*. Academic Press.

Chapter 12
Complexity Ramifications of High Stakes Examinations as a Measure of Accountability in Education Systems

Ssali Muhammadi Bisaso
Islamic University in Uganda, Uganda & Hacettepe University, Turkey

Shakira Bodio
Islamic University in Uganda, Uganda

ABSTRACT

Examinations occupy a strategic position in our lives and society today. In a situation where moral principles, rules and regulations for conducting examinations are truncated by either teacher or the learner, the validity and reliability of examination and certification are at risk. The nature and process of examination is something that should be looked into deeply having known the role of examination in our lives. That is essence of this paper; the way high-stakes examinations have affected various aspects of the education system in the realm of teaching, learning, assessment, discipline, roles of stakeholders, teachers' beliefs, community expectations, functional role of education, educational outcomes, individual differences, psychological status of learners, educational returns, school completion etc. This paper articulated these issues in terms of using high-stakes examinations as a measure of accountability in the education system and premised on the level of faith placed on these high-stakes examinations irrespective of educational value thereof.

DOI: 10.4018/978-1-5225-0460-3.ch012

INTRODUCTION

Education is a fulcrum and rightly so, since all other developmental facets of life are hinged on it (Denga & Denga, 1998 cited by Nweze, 2009). Nweze (2009) also asserts that, "education is an ongoing process that facilitates the process of learning and acquisition of theoretical and practical skills that equip mankind to utilize the environment effectively for personal as well as national development". Proper education, in this case, has to do with the "balanced development of the cognitive" (intellectual), affective (emotional/feeling, ethical, moral) a psychomotor (practical) domains of an individual since this will curtail the development of simply one aspect of the individual while ignoring the relevance of others.

Since theoretical and practical skills needed for both personal and national development, it remains important to find out the extent to which the learners have acquired the same skills. This is because education is all about accountability on the part of educational managers to the various stakeholders especially guardians and educational authorities. The only process through which this may be clearly ascertained is known to us as examination. Examination according to Adeyemi and Akindele (2002) is "the process, which comes after a period of learning, and it is an organized assessment of an individual's performance, on the basis of his or her institutional procedural exposure". Nsude (1998) looks at examinations as "tools meant for evaluation of the progress made by individuals in the course of acquiring skills or seeking knowledge over a specific period of time".

Examination, premised on the above, has to do with the "passing of value judgment on an individual", on the basis of the individual's performance in a set of questions, statements or series of tasks given, with the intention of assessing how much of a desired skill, trait, knowledge or behaviour the individual possesses. In some cases terms like measurement, assessment, testing and evaluation are used interchangeably to refer to examination.

The Oxford Advanced learner's dictionary of contemporary English defines the term examination as the "process or act of testing for knowledge and ability in order to determine its worth". The implication of this is that a test is a means of "measuring the knowledge, skills or aptitude of an individual". The word examination is indeed a broad term which connotes the assessment within an educational system meant to establish how much has been acquired by learners in terms of knowledge, skills and intelligence (Osindeinde, 2000 cited by Ibara, 2009). From this assertion, it is clear that measurement of the extent of learners' achievement in terms of educational objectives is the purpose and essence of examination. Even in distance learning institutions, examination serves a number of functions such as selection, certification, and accountability. But the fact remains that these examinations are indeed high stakes at all levels of measurement and evaluation.

High-stakes examinations are part of education policy design and scores on examinations are normally linked to grade promotion, high school graduation, and in many cases, teachers' salaries as well as their tenure decisions (Lewis, 2015; Wayne Au, 2007 citing Madaus, 1988). As part of the accountability movement, stakes are high because scores, and ranking and categorization of schools extending from the scores; are reported to the public (McNeil, 2000). In schools, examinations administered by various bodies are high-stakes because the scores determine school performance and thus, market status. Scores are also used to make important decisions that affect students.

At this level it is imperative to explain the term "high-stakes testing" as used in the context of writing and arguing issues in this paper by the authors. A high-stakes test is a test with important consequences for the one taking the test (Larra-Lee, 2011). Indeed when you pass, important benefits, such as a high school diploma, scholarship, or even a license to practice a particular profession. Meanwhile failure has important repercussions like being forced to go for remedial classes until that particular test can be passed, you may not be allowed to drive a car, or not being in position to find an ideal employment. In common usage, a high-stakes test houses major and serious consequences for the one taking it and is normally a basis of a major educational decision (Larra-Lee, 2011; Wikipedia).

Precisely therefore, a high-stakes test reflects any test that:

- Is a single, defined assessment,
- Has a clear line drawn between passing and failing, and
- Has direct and far-reaching consequences for passing or failing (there is something "at stake").

"High stakes testing" is something of a general term. In most cases, high stakes testing is taken to mean testing that has "serious consequences" for the individual taking the test (Greaney, 2004). Some examples of high stakes tests would be one where, "if you didn't pass, you wouldn't be allowed to advance to the next grade, or where you wouldn't be allowed to graduate or wouldn't get advanced placement" (Lewis, 2011; McNeil, 2000). The term "high stakes test" reflects most of the standardized national tests, which have a consequence regarding admissions, promotion or graduation.

In various parts of the world, secondary education sub-system largely operates on the assumption that high stakes examination scores provide reliable and objective evidence of school effectiveness (Lewis, 2015; McNeil, 2000). The same scores are used to detail student performance, the extent to which schools provide students with promised learning experience, and overall performance of the education system

(Greaney, 2004). The examination scores thus, represent the fundamental account that defines the presumed strength of the schools' learning environment, family-school relationship, and the outcomes that these relationships engender.

Relevance of Examinations to the Education System

The capacity to evolve over time determines good education systems. Testing systems are also expected to possess this capacity, both in relation to their purposes and the actual assessment instruments that they create (Nweze, 2009). Given the more rigorous demands on teaching as well as learning that have become internationally accepted, test validation requires a "concomitant rigor with a broad range of strong evidence". In principle, writes Sandra (2013) "teacher assessment has an important role to play in increasing assessment validity by complementing testing to cover subject domains more comprehensively than otherwise would be possible".

To achieve the corresponding educative value, high quality exams will require radical change in system design. Meanwhile, the importance of examination(s) can be seen from the following:

- Development of the spirit of hard work and zeal to study within candidates.
- The source of data and information for administrative action is simply examination.
- Affording authorities an opportunity to ascertain the extent to which knowledge has been successfully imparted.
- Examination serve the government as a means of educational control.
- Candidates become active participants within the teaching-cum-learning process.
- Students with special skills can be properly identified for placement and other decision making purposes.
- Effectiveness on the part of both teachers and students within the educational system is ensured by examinations.
- Candidates are selected for employment basing on examination results, this also aids educational advancement.
- The viability of a teaching method is established through examinations
- Examination results feed research undertakings.
- Examinations help in discovering learning problems and difficulties encountered by students.
- Guidance and counseling programs may also be premised on examination results.

The Challenges Vested within High Stakes Examinations

In recent years, the use and misuse of high-stakes tests has become a controversial topic in public education, especially in the Uganda where they are used not only to assess students but in attempts to increase teacher accountability and decision making within the education system.

The goal of evaluation is to make the examination system educative to all stakeholders within the education system: the series may reflect parents, teachers, students, as well as policy makers (Blazer, 2011; Lewis, 2015; & Nichols, 2012). Timeliness is central that is why Lewis (2015) laments that "if the exam results, scoring rubrics, and sample papers are not returned promptly, the window of interest, engagement, and learning for teachers, parents, and students will have closed. Teachers and students will have moved on to other parts of the curriculum and have no enthusiasm for feedback that is not currently relevant".

Premised on the above therefore, high quality assessment enjoys and plays an integral part in a coherent education system, and is thus linked directly to the improvement of the teaching and learning process as well as professional practice (Lewis, 2015). But since all complex adaptive systems are like this, the above should not surprise us at all: assessment provides us with feedback which is critical to enhancing every aspect of the system's performance. This however goes down as a strategic and technical design challenge. Being a technical challenge implies the need for technical expertise to deal with it at every level within the system.

Kirkpatrick and Zang (2011) argue that good teaching cannot be considered an effect of a test if a test is deficient due to what they term as "construct under representation or construct irrelevant characteristics", and conversely, if a test is faced with what they coin as "construct-validated", even if poor teaching would be involved, negative backwash cannot be largely associated with that particular test. According to Lewis (2015), Howie (2012) and Larra-Lee (2011), therefore, authenticity of the result of any examination can only be viable under the following conditions:

- Candidates sitting for the examination must have been taught well.
- No candidate should not be allowed to register and thereafter take an examination when their attendance is not up to 75%.
- Without valid and clear reason(s), lateness to any examination by all stakeholders ranging from invigilators or supervisors to candidates or students is inexcusable.
- Zero tolerance should be instituted for cases like leakage of examination questions, impersonation, cheating, collusion with other candidates or ex-

aminers or external agents, bringing in unauthorized materials, leaving an examination hall without authorization and many other aspects of examination dishonesty.

- The hall used for examination must obtain in the realm of conduciveness; with proper ventilation and illumination, sufficient chairs and tables for all candidates, non-threatening and friendly environment, properly spaced sitting arrangement, examination papers must be enough as well as answer scripts, and proper allocation of time, etc.
- The level of the candidates must be considered during construction of the test and the same must be based on what has taught by the examiner to the candidates, not just out of their syllabus experience as teachers. A matured personality is expected of an examiner in order to be considered responsible, etc.

It is the improper handling of assessment or examinations that has brought in what some may term "examination malpractice" or examination offences or "examination irregularities". Once examination malpractices take center-stage, then the credibility of high-stakes examinations is brought into disrepute. Against that background, Greaney, 2004; McNeil, 2000; Sandra, 2013 and Wayne Au, 2007 opine that High-stakes tests are often criticized due to the following issues:

- Failure of the test to measure the individual's knowledge or skills correctly.
- What the critic wants may not be measured by the test effectively.
- High-stakes testing can result in a narrow curriculum and lower skills. It encourage teachers to omit material that is not tested which is referred to as "Teaching to the test"
- Testing causes stress for some students. The basic term for this is test or performance anxiety.
- In many cases, high-stakes tests are given as a single long exam instead of continuous assessment test. Indeed some critics would choose a continuous assessment mode.
- High-stakes testing creates more incentive for cheating.
- A high-stakes test is sometimes tied to a controversial form of reward.
- Tests do penalize some test takers who do not have the required skills even in circumstances where it is no fault of their own.
- High-stakes tests may also reveal that some of the examinees do not know the necessary material, or do not have the required skills. As such, failure has a lot of consequences for such people especially in the public domain.

Discrepancies in the Education System Due to High Stakes Examinations

The education system the world over depends on examination results to deal with students: this encompasses admission, promotion, demotion, retention, graduation, ranking etc. Figlio and Lucas (2003) opine that while high standards have been advocated by policy-makers, business groups, and teacher unions, very little is known about their effects on outcomes.

In some analyses, high-stakes testing, grading and examination has been found to have a touching positive impact (Betts & Grogger, 2000; Blazer, 2011; Figlio & Lucas, 2003; Nichols, et al, 2012; Tugba, 2012). The claim is that, in sum, "the predicted effects of higher grading standards are small in terms of variation in achievement among students, but substantial in terms of the mean gain in student achievement and students respond favorably to the incentives provided by higher standards". Indeed Betts and Grogger (2000) argue that test scores rise is synonymous with higher standards schools, but even then, they rise more for students near the "top of the achievement distribution" than for students near the bottom. Equally, Blazer (2011) contends that although the majority of unintended consequences are negative, researchers have found that high-stakes tests have some positive effects on education, including:

- Increased professional development for teachers,
- Proper alignment of instruction with content standards for the state,
- Efficient remediation strategies for low-achieving students, and
- The use of data that informs instruction increases.

Greaney & Callaghan (1995) are however, uneasy about the absolute faith placed on high stakes examination scores as the key valid and reliable accountability measure. Such concerns according to them may be premised on the following:

- First, ranking schools basing on raw examination scores does not take into account the differences in school intake and the physical conditions of the schools.
- Second, no adjustments are made to examination performance on a value-added basis.
- Third, errors in measurement are seldom taken into account when judgment of merit is being made.

Furthermore, Cobbold (2009) and Nicholas & Berliner (2008, 2005) show that schools tend to manipulate high stakes examination scores by:

- Teaching to test,
- Narrowing the curriculum,
- Retaining students in untested grades,
- Eliminating low-performing students, and
- Outright examination fraud.

For a number of reasons, increasing reliance is being placed on teacher assessment in high-stakes contexts in many countries around the world (Greaney, 2004; Lewis, 2015; Sandra, 2013; Tugba, 2012). Simultaneously, countries that have for some time relied to greater or lesser degrees on teacher assessment for high-stakes purposes are in the process of questioning the validity of that reliance.

The critique of high-stakes examinations and the faith placed on them as a measure of accountability and competence in the education system can only be understood and conceptualized once considered from the perspective of major "idiocies within systems of education". Considered from a Ugandan realm, high-stakes examinations and their "untold effect" on the education system can therefore be viewed from the following perspective:

- **Credibility and Integrity of Schools:** Manipulations which verge on academic dishonesty cast doubt on the validity and reliability of examination scores in systems that use the scores as the fundamental account detailing; how much students learn, how well they can perform, and whether schools offer the promised learning experience (Amerin-Beardsley, 2010). Thus, the possibility that competitive pressure can lead schools to manipulate examination scores to improve measured performance challenges the premise that the market renders students the beneficiaries of a fruitful education experience.

Assessment behavior, which Betts & Grogger (2000) as well as Gredler (1999) and equally echoed by Jackson & Davis (2000) opine is in reference to the school conduct during the course of examinations, is an issue of critical debate in lieu of high stakes examinations. Whereas it is mandatory for schools to strictly observe the rules governing national examinations or zero tolerance to exam malpractice; the schools seek to win or not to lose by indulging in examination fraud.

Forms of examination fraud identified by Musisi (2012) include:

1. **Leakage:** Obtaining confidential examination content before it is officially released to schools.
2. **Contraband:** Encourage candidates to smuggle material into examination rooms with intention to cheat.

3. **Collusion:** Encourage candidates to assist one another during the course of the examination.
4. **Impersonation:** Facilitate another person to sit examinations on behalf of registered candidates.

Musisi (2012) contends that further fraudulent practices may also reflect the following:

1. **External Assistance:** Facilitating a more competent person to assist candidates answer examination questions correctly.
2. **Substitution:** Replace a candidate's original work from the script envelop with work done outside the examination room or outside the period of the examination.
3. **Peacekeepers:** UNEB-deployed invigilators or supervisors compromised by schools to let and/or help candidates cheat.
4. **Examination Attachés:** Examiners that connive to mark fellow conspirators' school scripts auspiciously.
5. **Redeemers:** UNEB examiners, computer operators, or checkers bribed to alter candidates' initial scores to make reported performance look impressive.

If and when schools do such, then "credibility of such schools within the public domain" remains doubtable: consequently, educational expectations keep dwindling to a point of no return indeed.

- **Authenticity of the Evaluation System:** Since some of those who handle teaching and learning are not professional teachers, knowledge of the ethics of the teaching profession eludes them significantly. Indeed Bisaso and Rafiyyah (2010) found that their perceptions of learning and teaching practices they put across obtain within a traditional paradigm. Sequel to the above is that implementation of the un-known is a daunting impossibility to everyone. Beardsley (2010) opines that "the pervading emphasis on grades and certificates in the society have made some examiners, parents and candidates to throw caution to the wind in a bid to have a high score for either to increase the reputation of the school or their candidates' chances of securing a good job in the labour market".

Those who set the exams themselves are not trusted that much which may lead to leakages in the examination process (Amerin-Beardsley, 2010). Given that the system lacks credible examiners leaves a lot to be desired when discussing the quality of the education system which bases its accountability claims on high stakes examinations. Otherwise with a normal system, everyone would be stress free and focus on developmental processes for his or her student within the education system.

- **Teachers' Beliefs about Education:** Teachers' practices are deeply rooted within their beliefs and actually influence all that a teacher does and plans to do (Bisaso & Rafiyyah, 2010). Teachers make conclusions about students depending on the quality of report cards they have meaning that a higher grade on a report card would make the student appear good in the eyes of the teacher, but the kind of information that can enable teachers to help students raise their level of performance cannot be obtained from report cards (Amerin-Beardsley, 2010). Fine students can receive a score and fine they can secure a placement in the class as a whole, but information that would be of help to them in terms of understanding what higher quality performance involves cannot be obtained.

Though teachers are aware of the relevance of including "active and student-centered strategies" on a regular and consistent basis, the state testing seems to drive the curriculum and warrant more "teacher-focused instructional methods, lecture worksheets and whole class discussion" in additional practice (Gredler, 1999; Jackson & Davis, 2000). What teachers need is detailed knowledge about their strengths and weaknesses, differential information on instructional ideas which all can help to remediate weaknesses. With high stakes testing however, such issues and ideals are largely impossible.

- **Classroom Teaching Practices:** It is widely believed that one of the most effective ways of creating change in an education system is by introducing or by re-designing high-stakes examinations (Musolleno, 2010; Wall, 2009; & White, 2010). This method is not foolproof, however, as there are many factors which can affect the impact of such an innovation.

With high stakes examinations or testing, "the intended outcome is altered by factors in the exam itself, as well as characteristics of the educational setting, the teachers and the learners" (Tugba, 2012). Such alterations include "examination impact and innovation in education". Accordingly there is need for guidelines to educators who continue to believe in the potential of examinations to effect curriculum change. So in this case, teacher teach to have learners passing the exam i.e. on-the-exam teaching mainly punctuated by endless revision and sample reading questions geared towards preparing learners for the examination, moreover with a high grade. This shifts the whole idea of teaching to coaching.

- **Honesty and Trust of Educators:** No one can deny the fact that there is acute poverty on the part of teachers, moreover they constitute the chunk of our key examiners and invigilators. Once this is agreed upon, then it is hard to deny the possibility of having the same teachers tamper with the credibility of the examinations (Beardsley, 2010; Amerin et al, 2010). The end result of this is the dwindling respect and trust given to teachers who deal with the assessment system.

Even those at the level of marking national examinations cannot be fully trusted, whilst those who set the examinations are doubted, and those who invigilate are risk-laden. Even at school level some teachers have found themselves either being compromised or sympathizing with students thus ending up offering a pass mark to get them through despite failing to raise the required grades.

- **Students' Conception of the Learning Environment:** Kelly (1991) defines a conception as a "mental image of reality containing beliefs, meanings, preferences, and attitudes that explain complex categories of experience". Research establishes a strong connection between conceptions of learning and how students approach learning (Brown, et al., 2008; Entwistle & McCune, 2004; Biggs, 1999). Differences in ways students approach learning has been attributed primarily to their conceptions of the learning environment and tasks.

Furthermore, Brown, et al. (2008) reiterates that the way teachers teach, the learning activities they assign to students, the workload students have to deal with, the assessment procedures used, and the degree of academic freedom given to students; exert the most significant influence on how students approach learning. This implies that "students' learning" primarily obtains within what they think their teachers expect them to do. Entwisle & Tait (1990) show that students that are constantly relying on surface approach to learning actively prefer teachers who provide pre-digested information. On the other hand, students with deep approach prefer teachers who assign them challenging and stimulating tasks.

Assessment exerts profound influence on how students learn. If assessment rewards understanding, learners will seek to understand the material taught; and if assessment rewards rote learning then that is what learners will do. With high stakes testing therefore, learners are expected to cram and commit information to memory for the sake of passing examinations after which they won't be able to even recall nor understand knowledge values.

- **Students' Approaches to Learning:** These are strategies students adopt to learn given tasks in a given learning environment. The traditional approaches to learning are; deep or understanding vis-à-vis surface or reproducing approaches to learning (Marton & Säljö, 1976 cited by Musisi, 2012). Later on, the strategic or achieving approach was developed by Biggs (1983). However, Richardson & Price (2003) point out that a strategic learner can be a deep or surface learner. Thus, this paper dwells on the negativity within surface learning.

This approach to learning constitutes the tendency of learners to handle learning tasks in a superficial way by adopting rote memorization techniques (Warren, 2004); focusing on other than quantity of information acquired (Tagg, 2003). The key aim of the learner is to cram material to pass the test or avoid failure other than seeking to understand how the material relates with other situations (Biggs, 2003). Surface learning is promoted by a curriculum jam-packed with content, teaching that presents material as bits of unrelated facts, allowing learners to be passive, emphasis on coverage over depth, assessment methods that emphasize factual recall, short assessment cycles, and low expectations of success (Atherton, 2005).

The problem with surface learners accordingly can be viewed from the analysis of Musisi (2012) thus:

1. Typically receive information unquestioningly,
2. Find difficulty to distinguish between principles and examples,
3. Treat different topics as isolated bits of content,
4. Find difficulty to recognize new material as part of previous material,
5. Convey disinterest or even a negative attitude towards learning,
6. Hold a cynical view about education,
7. Prefer non-academic areas such as socializing or sports,
8. Believe that factual recall is all that matters, and
9. Suffer anxiety attacks due to low expectations of success and fear of failure.

Whereas educators and other key stakeholders in the education system would seek to erode such prospects within the learners, the reality will remain that success in such an undertaking is impossible since the sole cause for most of these aspects is indeed high stakes examinations.

- **Classroom Achievement and Attainment:** Due to the pressure of "educational reform and high stakes assessment", some schools, risk foregoing active, and student-centered teaching and learning activities to give way to test taking skills and memorization of the discrete facts (Blazer, 2011; Gredler, 1999; Jackson & Davis, 2000). With additional pressure to fulfill the expectations of education under the auspices of "No child left Behind" (NCLB) educators are likely to feel like abandoning the tenets of school philosophy for more teacher-centered instructional approaches. Since such policies were enacted, teachers have been accountable to students' achievement and implementation of the state assessment standard. Thus higher standards may contribute to "greater inequality in the distribution of educational achievement" (Betts & Grogger, 2000)

Issuing from the above premise, therefore, "students with higher potential" may continue posting high scores on tests and exams while those with a low potential are likely to be crashed by the system and end up attaining poor results. Since education would ideally be the centre of establishment of equitable distribution of learning opportunities, high stakes examinations would, in this regard, render the education system a centre of discrimination and stratification among learners.

- **Learning Outcomes:** According to Becta (2009), learning outcomes constitute what the learner is expected to know, understand, and be able to do on completion of the learning process; evident after learning has taken place. Learning outcomes can be evaluated quantitatively or qualitatively; depending on whether learners focus on reproducing learning material or on seeking meaning and application of the material (Tan Po Li, 2005). But owing to the wide-ranging requirements that need to be met by secondary school graduates; stakeholders define learning outcomes in general terms. However, whatever the descriptions given; the general consensus is that learning outcomes should be functional, durable, flexible, meaningful, generalized and application-orientated.

With high stakes examinations therefore, it becomes hard for learners to know, understand and be able to do anything especially after the examination. There won't be anything called learning outcome but simply examination outcome. They will know everything virtually at the time and during the examination but largely nothing at the end of the examination day.

- **Functional Role of Education:** According to Bregman & Bryner (2003), education prepares students for lifelong education, and for the world of work. Education also develops vocation-oriented skills, prepares children to participate fully in society, to take control of their own lives, to continue to learn, and provides a platform for transition from childhood to adolescence (Howie, 2012; Nichols, 2012). Based on the foregone, school systems need a valid, reliable and clear picture of students' performance, as well as that of classes and the schools. This calls for validity which requires tests that are clearly balanced across the performance goals, which will solve the testing jig-saw by abrogating easy testing or target testing.

One aspect of such a system condemned by Robinson (2008) is high stakes examinations as a measure of accountability. In this system only those who best "conform to the standards and values of the more total ideological scheme rise through the hierarchy of academic evaluation". That implies that at the end of the day, education will only represent those who can score a higher mark on a test or examination and anything short of that will not count. Tellingly then, the pillars of learning (UNESCO, 2008) would have lost value and no one in the school and education system would be even interested. So education and schools will seize being centres for preparation of students for life to becoming centres for preparation of students for high stakes examinations.

In a broader sense, exam-centric means a system in which control is instituted on what students know and do not know. It thus manifests as an oppressive force sometimes, to one's academic as well as social lives. Critical thinking, which is an ultimate purpose of education is simply downplayed in an education system that cherishes high-stakes testing as a core component.

- **Admission Behavior of Schools:** Betts & Grogger (2000) contend that admission behavior constitutes the criteria used by schools to admit applicants expected to give the school competitive advantage. The criteria include *Brain power*; targeting academically brilliant applicants whose scores are expected to elevate the school ratings on examination league tables.

With the need for high scores being the ruling policy of admission, the end result would be leaving out those students who do not seem good enough in the eyes of the school administrators. Such learners would not only be deprived of the opportunity to attend good schools but also the opportunity to compete with the best in order to improve. Most crucially though would be the issues of educational equity (UNESCO, 2008) as well as education for all (UNDP, 2004). The practice is that the best schools in the country sit and select the best candidates from the national examinations leaving the resource constrained schools with the less performing candidates, the cycle thus becomes uninterrupted in the end.

- **Retention Behavior of Schools:** Betts & Grogger (2000) contend that keeping students in schools is a critical undertaking for all stake holders in the education system. However with high stakes examinations taking center-stage, the retention behavior of schools has drastically changed. This is in reference to the criteria the schools use to deal with learners of diverse learning abilities so as to boost a school's competitive advantage.

With high stakes testing, the criteria include: Cataloguing; assigning students to streams basing on academic ability and cleansing based on eliminating students whose academic performance is likely put the school in academic disrepute (Musisi, 2012; Nichols, 2012). Once such becomes normal school practice, chances are high that some innocent students are likely to be kicked out of the schools which do not want to risk their reputation and integrity. In some cases, schools have registered their candidates in other lower caliber schools once deemed less fit to post incredible scores. Crucially, sometimes parents of the children are not even aware of such acts of treachery on the part of schools.

- **Students' Psychological Settings:** Kirkpatrick and Zang (2011) observe that an "exam-centric education" causes extreme stress for students since tests are regarded as the only means of proving their worth. This becomes a filtering process that selects a few of the students for higher education. An "exam-centric education" assigns student aptitude to test scores, oftentimes leads to marginalization, if not outright disregard, of lower performance students (Banks et al, 2015; Blazer, 2011; Mitchell, 2012; Reddell, 2010). In Uganda, lately, just like in China and other parts of the world, worth is not defined by people but by scores. Even parents and teachers have been influenced to start considering scores as the only criteria for evaluating students. Therefore, high scoring students garner praise whereas low or mediocre scorers are liable to punishment.

The result of this "exam-oriented education" is that a large number of weary students will be left with inadequate psychological development, coupled with repressed personality, self-hatred, and a general lack of motivation in the development of other abilities. This type of an education system is only fit for machines not humans since there is a social aspect attached to Human beings unlike machines, and this should not be neglected.

- **Individualized Learning Styles:** From audio to visual, and from audio-visual to kinaesthetic, learners' learning styles have always and will always differ. Some children have a performance for learning by sight, for example. They are called "visual learners". Others who use hearing as the preferred sense are known as "auditory learners". Still others who prefer touch or whole –body involvement in their learning are called "kinaesthetic learners" (Sousa, 2001).

An "exam-oriented education" system not only increases a students' burden but also "restricts a student's ability to learn using techniques that a particular student finds most effective" (Kirkpatrick & Yang, 2011 and the incentive to learn is largely killed in this case (Blazer, 2011; Betts, 1995; Larra-Lee, 2011). With high stakes exams, students will not have the audacity to do learning their way or better still even be able to adopt and develop a learning style, what is at stake in this case is grades or scores from the exam and the most effective technique here normally turns out to be cram work, a highly castigated practice indeed.

- **Students' Discipline:** Figlio and Lucas (2003) seem to claim that high stakes examinations transform the behavioural settings of learners in a school system by making them steadfast and focused on the task. Equally, Betts and Grogger (2000) posit that with high stakes examinations, learners are given a clear lead and focus which helps in improving on their commitment to the school cause and keeps their discipline in check.

Notwithstanding the above however, Kirkpatrick and Zang (2011) are uneasy about such views since they claim that with high stakes examinations, cheating becomes a common phenomenon. The view is also extended by Amerin-Beardsley (2010). High school students are over-burdened with homework on a daily basis that exceeds their "normal capacities to memorize" and in a bid to pass exams, their only alternative turns out to be cheating, this is not only a common thing in secondary schools, but also universities and in day to day life as a whole.

The educational system has become too competitive and thus demands high test scores over and above anything else, so, students' reaction to this has also been spontaneous: they attempt cheating on examinations to mitigate the risk of failing. With too many tests, students get no time to digest knowledge, and as a way of escaping punishments, they choose cheating in order to post high grades, possibly as others. Since one success is known to drive another, it is taken on even at the risk of it being a cheating undertaking.

- **Individual Differences among Learners: S**tandardized tests are the worst means of assessing the abilities of students with learning deficiencies whether with accommodations or not. Given the individual learning styles of students with learning deficiency, it is felt that these high stakes tests pose considerable danger to a student (Blazer, 2011; Banks, 2015; Larra-Lee, 2011; Reddell, 2010). Often, standardized tests are poorly devised and give very little consideration to the validity of the assessment they provide to learning deficiency-laden students.

Secondly, according to Yell, Mitchell, Katsiyannis, Antonis, and Losinski (2012) these students with disabilities "often perform poorly on these tests when compared to the mainstream population" due to a few issues which include terrible neglect on the part of school teachers and administrators and the belief that these students may actually affect the average of grades posted by a school. It is not surprising therefore, that sometimes these students are even encouraged to leave the examination/ this kills self-esteem and abrogates self-confidence on the part of these learners.

- **Educational Attainment and School Completion Rates:** Betts & Grogger (2000) put it that, "in those states that raised graduation requirements, drop-out rates did not decline as much as in other states over the same period". The estimates above would suggest that higher standards may lead to "slight decreases in high school graduation". Nevertheless, the attendance levels in schools keep dropping due to high grading tendencies (Blazer, 2011; Howie, 2012; Nichols, et al, 2012).

In many cases, by the time students complete a particular level of education, a sizable number would have dropped out. Often a time, educators, researchers, policy makers and the entire bunch of stake holders point to varying factors that explain and justify their credibility in their respective domains. However, there is no gain saying that these dropout rates are normally due to the pressure levied on them by the education system to obtain high scores week in, week out. Students who cannot stomach or shoulder such burdens and pressure brought to bear on them by a high stakes examination-laden education system would find quitting a safer route.

- **Educational Returns:** Recent research has shown that the higher the test scores the higher the wages and thus students with lower scores find themselves less competitive in the labor market where payoff to finishing high school as well as attending college is found to be even greater (Betts & Grogger, 2000; Figlio & Lucas, 2003). Now that is the scenario with our education system where every aspect of life is judged upon grades.

As if it is not enough to fail to get to a school due to the requirement for higher grades by that particular school, it becomes even more daunting to stay in the school since the retention behaviour of the school will equally be determined by high grades. Tellingly more annoying and perhaps even unrealistic is the fact that earning a good job and pay after completion of education is also premised on excellent grades (Blazer, 2011; Larra-Lee et al, 2011; Reddell, 2010). The viscous cycle thus continues to take shape where the poor remain poor at all levels and the rich remain so accordingly.

FUTURE TRENDS

With evaluation becoming increasingly more and more challenging, yet a central theme of our education systems the world over; Judging a student's performance on the basis of multiple reviews stands out as critical to assessments and evaluations. Current practices focus solely on a learner's ability to recall facts which does little to effect holistic development. There must be a paradigm shift in evaluation systems.

Conducting regular peer evaluations of teaching, even for senior staff. This would help in developing competences at all levels that would aid effective evaluation decisions in the education system. Sometimes the challenge lies in novices partaking in key evaluation systems.

Reviewing syllabi and learning outcomes. Sometimes evaluation is based on unrealistic learning outcomes thus leading to misplaced results in the education system. Focus in education systems should be premised on effective learning outcomes.

Creating a variety of evaluations to supplement (but not replace) the standardized form. Sometimes the prevalent system of evaluation is so rigid and cannot allow flexibility to prevail in the system. So there is need to consider other forms of evaluation to aid the established one. Research on the applicability of such ideas in the field of evaluation ought to take centre stage at all levels in the education system.

CONCLUSION

Examinations are paramount to our educational, social, political, economic and other life systems today. They are a benchmark for success in all spheres of influence and they form the core of giving a true picture of performance. They are also largely agreed upon and trusted by many stakeholders in the education especially parents, teachers, counsellors, administrators and policy makers. Even students themselves have learned how to respect and trust examination results since they depict who is better than who in the school system. Parents also wait for examination results to know how well or otherwise their children are performing. Counsellors look out for poor performance to swing into action and teachers make reports based on scores from tests. When done, the school administrators win bragging rights when their schools top the performance pyramid while policy makers would also use the scores to rank the schools.

With the above in place, examinations ought to be very good and properly managed so as to serve the seemingly sacred role brought to bear on them by the oft unforgiving education system. Every effort should thus be made in order to harness every resource "both human and material" needed in order to achievement of the objective(s) of education in the school system. Thus in keeping with Kirkpatrick and Zang (2011), this paper suggests "moderate pressure to excel in school", when paired with deemphasizing high-stakes testing. This will better motivate students and improve student success as well as psychological health, both in terms of "academic success" and also a "productive adulthood".

Uganda's approach to education calls for acute reform if its relevancy is to be maintained. The following serve as key recommendations that all educators and stakeholders should give a proper thought, consideration and action:

- More attention to the teenagers overall harmonious development should be conceived by educators.
- Opening up the education system to flexible examination strategies should be made gradually.
- Pilot schools should be opened up.
- Running of schools should be in a non-bureaucratic manner.
- Healthy competition based on skills and competencies ought to be instituted in schools.
- Promoting reform in our traditional schools in order to re-invent the wheel.

This will usher in competition between schools: ultimately improving on performance as well as overall quality. Examinations are indeed a desirable thing, but when they are high stakes, they lose meaning since the cons outweigh the pros and by a long shot.

REFERENCES

Abraham, R. R., Vinod, P., Kamath, M. G., Asha, K., & Ramnarayan, K. (2008). Learning approaches of undergraduate medical students to physiology in a non PBL and partially PBL- oriented course. *Advances in Physiology Education*, *32*(1), 35–37. doi:10.1152/advan.00063.2007 PMID:18334566

Amrein-Beardsley, A., Berliner, D. C., & Rideau, S. (2010). Cheating in the first, second, and third degree: Educators' responses to high-stakes testing. *Education Policy Analysis Archives*, *18*(14), 36.

Atherton, J. S. (2005). Learning and teaching: Learning how to learn. Academic Press.

Atherton, J. S. (2008). Doceo: Teacher-centred teaching works. Academic Press.

Banks, J., & Smyth, E. (2015). Your whole life depends on it: Academic stress and high-stakes testing in Ireland. *Journal of Youth Studies*, *18*(5), 598–616. doi:10.1080/13676261.2014.992317

Becta. (2009). *Literature Review: Deep learning with technology in 14- to 19-year-old learners.* Retrieved from http://publications.becta.org.uk/display.cfm?resID=31532&page=1835

Betts, J. R. (1995). *Do grading standards affect the incentive to learn? Manuscript, University of California.* San Diego, CA: Department of Economics.

Betts, J. R., & Grogger, J. (2000). *The impact of grading standards on student achievement. Educational attainment and entry-level earnings.* NBER Working Paper 7875.

Biggs, J. (1987). *Student approaches to learning and studying.* Hawthorn, Australia: Australian Council for Educational Research.

Bisaso, S. M., & Rafiyyah, S. A. (2010). *The relationship between perception of learning and teaching practices of lecturers: A study of Educators at IIUM.* Malaysian Research Repository.

Blazer, C. (2011). Unintended consequences of high-stakes testing. Miami-Dade County Public Schools.

Bregman, J., & Bryner, K. (2003). *Quality of secondary education in Africa (SEIA). Working African Development Bank, (AfDB).* Human Development Indicators.

Brown, G. T. L. (2002). Student beliefs about learning: New Zealand students. *Academic Exchange Quarterly, 6*(1), 110–114.

Brown, G. T. L., Lake, R., & Matters, G. (2008). New Zealand and Queensland teachers' conceptions of learning: Transforming more than reproducing. *Australian Journal of Educational & Developmental Psychology., 8*, 1–14.

Cobbold, T. (2009). *Local area school comparisons are 'simplistic' and unfair league tables.* Academic Press.

Entwistle, N., & McCune, V. (2004). The conceptual bases of study strategy inventories. *Educational Psychology Review, 16*(4), 325–345. doi:10.1007/s10648-004-0003-0

Figlio, D. N., & Lucas, M. E. (2003).*Do High Grading Standards Affect Student Performance?* University of Florida and National Bureau of Economic Research.

Greaney, V., & Kallaghan, J. (1995). *Equity Issues in public examinations in developing countries.* World Bank Technical Paper No. 272.

Howie, S. (2012). High-Stakes testing in South Africa: Friend or foe? *Assessment in Education: Principles, Policy & Practice, 19*(1), 81–98. doi:10.1080/096959 4X.2011.613369

Ibara, E.C. (2008). Administering examinations for quality control in distance education: The National Open University of Nigeria perspective. *Turkish Online Journal of Distance Education, 9*(3).

Kearns, L. (2011). High-stakes standardized testing & marginalized youth: An examination of the ımpact on those who fail. *Canadian Journal of Education, 34*(2), 112–130.

Kellaghan, T., & Greaney, V. (2004). *Monitoring performance: Assessment and examinations in Africa.* Washington, DC: World Bank.

Kelly, K. F. (2001). *Meeting needs and making profits: The rise of for- profit degree- granting institutions. ECS Issue paper*. Denver, CO: Education Commission of the States.

Kirkpatrick, R., & Zang, Y. (2011). The negative ınfluences of exam-oriented education on chinese high school students: Backwash from classroom to child. *Language Testing in Asia, 1*(3), 3. doi:10.1186/2229-0443-1-3-36

Lewin, K. M. (2003a). *Investigating the missing link—The case for expanded secondary schooling in Sub-Saharan Africa*. Academic Press.

Lewis, S., & Hardy, I. (2015, April3). Funding, reputation and targets: The discursive logics of high-stakes testing. *Cambridge Journal of Education, 45*(2), 245–264. do i:10.1080/0305764X.2014.936826

Madaus, G., & Clarke, M. (2001). The adverse impact of high-stakes testing on minority students: Evidence from one hundred years of test data. In G. Orfield & M. L. Kornhaber (Eds.), *Raising standards or raising barriers? Inequality and high-stakes testing in public education*. New York: The Century Foundation Press.

Marton, F., & Säljö, R. (1984). Approaches to learning. In F. Marton, D. Hounsell, & N. Entwistle (Eds.), *The Experience of Learning*. Edinburgh, UK: Scottish Academic Press.

Marton, F., & Saljo, R. (1997). Approaches to learning. In F. Marton, D. Hounsell, & N. Entwistle (Eds.), *The experience of learning* (pp. 3–22). Edinburgh, UK: Scottish Academic Press.

Musisi, B. (2012). *Pedagogic Ramifications of the competitive Secondary Education market in Uganda: A case of Proprietary schools*. (Unpublished PhD Thesis). School of Education, Makerere University, Kampala.

Musoleno, R. R., & White, G. P. (2010). Influences of high-stakes testing on middle school mission and practice. RMLE Online. *Research in Middle Level Education*, *34*(3), 1–10.

Nichols, S. L., Glass, G. V., & Berliner, D. C. (2012). High-stakes testing and student achievement: *Updated Analyses with NAEP Data. Education Policy Analysis Archives*, *20*(20), 35.

Nweze, T. (2009). Management of examinations: Ethical issues. *Edo Journal of Counselling*, *2*, 1.

Reddell, S. (2010). *High stakes testing: Our children at risk*. Online Submission.

Reid, W. A., Duvall, E., & Evans, P. (2007). Relationship between assessment results and approaches to learning in year two medical students. *Medical Education*, *41*(8), 754–762. doi:10.1111/j.1365-2923.2007.02801.x PMID:17661883

Richardson, J. T. E. (2003). Approaches to studying and perceptions of academic quality in a short web-based course. *British Journal of Educational Technology*, *34*(4), 433–442. doi:10.1111/1467-8535.00340

Robinson, K. (2008). *Changing education paradigms*. RSA.

Sandra, J. (2013). On the reliability of high-stakes teacher assessment. *Research Papers in Education*, *28*, 1.

Sousa, D. A. (2001). How the brain learns (2nd ed.). Academic Press.

Tugba, A. (2012). *An exploration of the effects of high-stakes examinations on the teaching and learning of mathematics in post-primary education in Ireland and Turkey*. (PhD thesis). National University of Ireland, Maynooth, Ireland.

UNAIDS. (2000). HIV/AIDS and the Education Sector. UNAIDS.

Warren, H. (2004). *Deep and surface approaches to learning: Learning and teaching theory for engineering academics. engineering subject centre guide*. Loughborough: HEA.

Yell, M. L., Katsiyannis, A., Collins, J. C., & Losinski, M. (2012). Exit exams, high-stakes testing, and students with disabilities: A persistent challenge. *Intervention in School and Clinic*, *48*(1), 60–64. doi:10.1177/1053451212449740

KEY TERMS AND DEFINITIONS

Accountability: The definition of accountability is taking or being assigned responsibility for something that you have done or something you are supposed to do.

Chaos: Wikipedia defines the term chaos as any state of confusion or disorder while at dictionary.com it is defined as a state of utter confusion or disorder; a total lack of organization or order.

Complexity: The state of being complex or difficult with a variety of challenges and obstacles to effectiveness and efficiency. Complex situations are born of lack of a clear approach to dealing with an existing situation.

Examination: This reflects the passing of value judgment on an individual on the basis of the individual's performance in a set of questions, statements or series of tasks given, with the intention of assessing how much of a desired skill, trait, knowledge or behaviour the individual possesses.

High-Stakes Test: According to Wikipedia, a High-stakes test houses major and serious consequences for the one taking it and is normally a basis of a major educational decision.

Chapter 13

Conflict Management in Higher Educational Institutions:
A Complexity Perspective

Farooq Miiro
Islamic University in Uganda, Uganda

ABSTRACT

Conflicts are inevitable and very common in educational institutions the world over. Conflict happens whenever and wherever in HEIs. Conflicts also rampantly occur at HEIs. Conflicts are therefore inevitable although this should not cause alarm truth is that they are natural and have benefits to organizational progress. It can therefore be noted that conflict management is part and parcel of school administration challenges that HEIs managers should handle as part of institutional politics, with extra care to benefit system and its workers. Therefore, experienced managers should motivate different conflicting groups to improve their methods and attitude towards work so that organizational goal can be achieved in the most effective manner. This however is a tough undertaking since it requires technical knowledge and skills on the part of leader. Accordingly, understanding complexity theory and applying it effectively in organizational management can transform conflict into an effective catalyst for improving school system.

DOI: 10.4018/978-1-5225-0460-3.ch013

INTRODUCTION

Today, the world is experiencing mass global increase in terms of population, heavy demands and challenges that require improved standards of living come into play as a result of technological advancement (Leung Yu, 2010), Education is also turning into a commodity with the aim of addressing global demands that call for critical thinking and problem solving (the ability to ask the right questions), collaboration across networks and leading by influence, agility and adaptability, initiative and entrepreneurialism, accessing and analyzing information, effective written and oral communication, curiosity and imagination (Wanger, 2008). It can be observed that the education system the world is adopting today, encounters serious challenges that range from social, economic and political within educational institutions.

As many education institutions are mushrooming at all levels, they attract staff from different education backgrounds, culture, climate and ethnicity to serve the common goal of the institution with emphasis on cooperation, harmony and collaboration that address the institutional culture and climate, although little attention is put on competition and conflicts. Through employee efforts to meet the common purpose that entails school believes, values and norms conflicts emerge in different types that require complex managerial approaches to administer them. As educational management literature has been in place for quite long, it has not been dissecting the dynamics required in dealing with HEIs conflicts. In the recent past some studies have been done to establish the management of conflicts in secondary schools but little has been written about the conflicts in HEIs. Therefore this conceptual paper would like to illustrate the complexity in the management of conflicts in Higher education Institutions.

In the context of HEIs, conflicts have been in place for more than 200 years ago, the first of its kind was in the University of Glasgow where the founder turned against his university in 18th C, within the same university another staff set the computer laboratory ablaze, in America's leading public university, Berkeley was distracted for five years to 2003, another case that merits closes study, however, is one which has the distinct advantage of being a matter of public record since the findings of the enquiry into the 'Great Battle' was published as 'The Davies Report' (1994). The conflict was centered on the Philosophy Department of University College Swansea and ran from October 1989 until May 1993, (OECD).

It can therefore be observed that as HEIs are viewing themselves as modern organizations to achieve corporate goals with the aim of addressing the dynamics of the changing world, conflicts are also created in modern society (Adepoju & Safowara, 2012). Conflicts are natural phenomenon and behaviors that exist among people through communication process as the internal or external discord that results from differences in ideas, values, or feelings between two or more people (Marquis & Huston,2000).

Conflicts are natural and unavoidable (Miklas & Kleiner, 2003), these are caused by different Several factors (both endogenous and exogenous) as identified by Nkomo (1984) that range from individual, group, students and staff and leadership management style, with multiple groups each with its interests HEIs are bound to chaos (Buchanan and Devletoglou, 1971). Some studies suggested that contradiction in goals between workers and organization can cause stress and then lead to conflict (Adepoju & Safowara, 2012). As staff respond to multiple tasks and demands from coworkers and students sometimes stress becomes inevitable (Avlonitis & Panagopoulos, 2006; Singh, Goolsby & Rhoads, 1994) and stressful working conditions can result in negative attitudinal and behavioral responses, and strain (Netemeyer, Maxham, & Pullig 2005). Stress can create conflicts due to incompatibility in job demands, un prescribed jobs which arises from unplanned responsibilities. Students' protests is to a large extent an expression of the social and political extremism which come through institutional rule and regulations yet they are not captured in constitutional laws governing the state (Adepoju & Safowara, 2012).

However democratic and tolerant the society is, there is no administration that can accept such behavior hence leading to conflict between students and administration. Buchanan and Devletoglou (1971) and Adepoju (2003a) citing Nixon, former American President, in one of his speeches stated; Intimidation and threats remain outlaw weapons in a free society. A fundamental governing principle of any great university is that the rule of reason and not the rule of force prevail. Whoever rejects that principle forfeits his right to be a member of the academic community. The university administrator who fails to uphold that principle jeopardizes one of the central pillars of his institution". As staff and students try to negotiate their way with administration to achieve their rights inform of wages and other motivating incentives, chaos arises when promises are delayed.

Major Causes of Conflict in Organizations

Most of the conflicts in organizations stem from Role ambiguity and work load related issues. Role ambiguity can cause friction between the supervisor and subordinate where by one of the two can look at the other as if is usurping powers or doing work that is not in his docket (Kahn et al. 1964; Rizo, House & Lirtzman 1970) with the nature of HEIs, run full-time, extramural, evening and external courses overload due to un employment and job turn over (Virick, Lilly & Casper, 2007).

Workload can lead to increased tension in stress where by sometimes managers can be looked at as if they are witch hunting some people within the organization (Abakus, Yavas, & Ashill 2009; Tucker et al. 2009). In a study carried out on some of the Japanese organizations, findings show that working to death and suicide overworking were the major problems faced by workers in the recent past (Kanai 2009). With this kind of life in the HEIs some staff term it as uncertainty and lack of appreciation for their future contract from the supervisor (Spector & Jex 1998). However genuine supervisors and subordinates enjoy doing such work because it attracts a high degree of experience (McKay & Tate 1999; Mulki, Lassk, & Jaramillo 2008; Narayanan, Menon & Spector, 1999). Role conflict can still be caused by the performance of an individual whereby if an employee does not execute his duties as required tension will be created on both the side of the worker and administrator calling for enforcement of rules and regulation which is always complex in nature due to unnecessary pressure (Owens & Valesky, 2011). In addition conflicts arise when interests collide usually as a result of divergence in organizational goals, personal ambition, group loyalties, demands on scarce resources and so forth (Rahim, 1986).

It can further be noted there other factors that cause conflicts when managers are improving on organizational performance, for example resistance to change due to transfers (Miiro, 2012), Foley (2001), in a time where many HEIs support collaboration, conflicts are bound to happen (Foley,2001).

Adepoju (2003) in his study about Nigerian universities cites the following as major causes of conflict:

- Quality of leadership,
- Character of economic environment,
- Character and posture of government in power,
- Government policies;
- Labour market,
- Faulty negotiations,
- Management, and
- Group behavior.

Types of Conflicts in Organizational Management and their Effect

Conflict is inevitable whenever two or more people interact, whether in the workplace or at home. Conflict can occur between two or more individuals, two or more groups, or an individual and a group. When dealing with conflict in an organization, it is important to remember to address the issue, not the people. According to

Rahim (2001) the key types of conflict that can occur in any organization include unclear definitions of role responsibility, conflict of interest, lack of resources and interpersonal relationships within the workplace.

- **Unclear Definition of Responsibility:** When it is unclear who is responsible for what area of a project or task, conflict can occur. Territorial issues arise when decisions are made that appear to cross boundaries of responsibility. To prevent this from happening it is imperative that the roles and responsibilities of all the players are spelled out clearly and agreed upon by everyone involved before the project is started.
- **Conflict of Interest:** Understanding how personal interests and goals fit within the structure of the organization will alleviate conflict of interest problems. When an individual's personal goals are at odds with the goals of the organization, the individual may be tempted to fight for his personal goals, creating a conflict situation that will hamper success of the project.
- **Insufficient Resources:** Competition for resources, including money, time and materials, will cause the teams to undercut each other, leading to conflict between departments or other work groups. Valuable resources need to be protected, as well as distributed fairly among all the groups. Starting out a project with a clear picture of the resources available will help waylay some of this conflict.
- **Interpersonal Relationships:** The personalities of the people involved in the organizational structure play an important part in conflict resolution. Often the conflict is a result of interpersonal relationships where the parties to the conflict are unable to resolve personal issues with each other. It is not always easy to set aside personal prejudices when entering the workplace, but it is important to recognize what those prejudices are and deal with them before conflict arises.

Meanwhile Nastase (2007) contends that conflicts can also take another form which features the following levels:

1. Conflict within the individual (intrapersonal),
2. Interpersonal Conflict,
3. Conflict between the individual and the group,
4. Intergroup conflict,
5. Inter-organizational conflict.

Conflicts can have both positive and negative effects of the performance of an organization. Conflicts have correlation between reduction in labour market crisis and productivity where education institutions cannot be exempted (Adepaju, 2003), they can be both psychological and physical withdraw. In his study of effectiveness in schools Msila (2011) states that the challenge of school principals is to identify the problems endemic in their school, he further cited Bola who stated that when the work of the administrator becomes unimaginable he fails to have time for school to run other business. It can jeopardize the performance of both lecturer and their heads of department (Msila, 2012).

When an environment turns unfavorable for workers due to an imaginable reason, it raises emotions between individuals (Covette, 2007). Buchanan and Devletoglou (1971) also described universities as institutions where unrepresentative but well-organized groups of students participate in "sit-ins", "invasions" or other activities that lead to upsetting the learning process. It is also described as a place where wanton physical destruction of property proceeds unabated.

HEIs are organizations whose future is unpredictable, whereby any day which passes comes with new demands of the time. Peters (1994) opined of "crazy organizations" that engulf the 21st century managers, therefore managers of such institutions need to alleviate fear from their staff to accommodate the pressures of the dynamic world. (Kotter 1999). Achieving success requires a manger to avoid hostility, laziness, and aggressive behaviors but all this is complex in nature whereby whenever the head stretches his leg he may step on other peoples toes. It can further be observed that when conflicts are not handled well, they may cause break down of school especially when the affected department is full of mammoth of tasks (Msali, 2012). It further stated that when teachers are involved into conflicts the challenges on behavior are indeed very disastrous in that supervisor tend to use harsh methods to curb them down and the whole situation turns into hostile (Corvette 2007, Croswel,2011)

Management of Conflicts in Organizations

Conflict is to be found in every organization. 'How well do you deal with conflict situations? Is the question to senior managerial post when they are arise except if they are contained, the institutional business may stay put. Every scholar within the university setting would wish to see a free blood institution for pride however the situation of conflicts is unavoidable due to natural factors. However the way managers of educational institutions handle conflicts to make education a clean place turns them into bad or good or functional and dysfunctional but this process is complex. However several authors suggest possible methods of handling conflicts.

Marquis and Huston (2000) identify five strategies that leadership of higher education may use in conflict resolution;

- Compromising or lose situation,
- Competing,
- Cooperating-Accommodating-Smoothing,
- Avoiding, and
- Collaboration.

Beebe, and Redmond (1999) identified five steps to facilitating problem solving:

- Define the problem,
- Analyze the problem,
- Determine the goals,
- Generate many solutions,
- Select the best solution, and
- Try the best alternative.

Blake and Mouton (1964), identified five conflict solving strategies thus:

- Smoothing,
- Compromising,
- Forcing,
- Withdraw, and
- Problem solving.

Furthermore, Dana (2001) states that leaders need to comprehend structure to be able to analyse conflicts well, this requires ability and skills from the mediate of the affected stakeholder (Mintzberg,1973), the issue of gender is also important complex matter to consider due to its sensitivity to avoid marginalization on any decisions may be take however research has always not considered it well (Rowleyet al., 2010), studies further indicated that women have been eliminated in higher position due to that fact that they may fail to handle conflict scenarios within educational institutions (Korabik & Ayman, 1989). In addition the element of cultural is influence is another complex fragile factor that should be handled carefully (Weiner, 2005). Anderson et al. (1996) considers negotiation as one of the most common techniques managers should adopt for creating agreements during conflict situations.

Another important factor is the organizational culture, while handling conflicts within an educational setting, they must fit within the organizational norms (Burke, 2004) in a study conducted by Gabrielidis et al. (1997) found out that gender and culture had an influence on the management of conflicts. Runde and Flanagan (2008) assert that conflict is at the root of many leaders' best ideas as well as the core of their worst failures. Therefore leaders need to learn from their mistakes to avoid unnecessary conflicts (Msila, 2012). However this kind of approach in most cases is complex to fulfill among leaders. As conflicts do not happen between staff only, sometimes they are caused by students. With this kind of situation in HEIS, Cole (1999) states that disruption and students violence in whatever way should be prevented because they have adverse results on organizational productivity.

CONCLUSION

Premised on information above, educational administrators are encouraged to treat conflict with care due to its influence that it can be felt within the patterns of behavior in the organizational setting (Hanson, 1991) due to that fact business in HEIs comes from faculties and the way its manned depends on how well managerial and faculty values become intertwined and then expressed in daily operating procedures' (Clark, 1998).

REFERENCES

Adepoju, T. L. (2003a). *Trade unionism and job productivity in human organization*. Ibadan: Hope Publications.

Adepoju, T. L. (2003b). *Trade unionism and culture of violence and resistance: A social system perspective*. Ibadan: Hope Publications.

Anderson, J. W., Foster-Kehn, M., & McKinney, B. C. (1996). *Communication skills for surviving conflicts at work*. New Jersey: Hampton Press.

Anderson, S., & Wahlstrom, K. (2004). *How leadership influences student learning*. New York: The Wallace Foundation.

Buchanan, J. M., & Devletiglou, N. E. (1971). *Academia in anarchy: An economic diagnosis*. London: Tom Stacey Ltd.

Clark, B. (1998). Creating entrepreneurial universities: organizational pathways of transformation. Perganon, UK: IAU Press.

Cole, G. A. (1999). *Management: Theory and practice (5th ed.)*. London: Letts Educational Aldine Place.

Corvette, B. A. B. (2007). *Conflict management: A practical guide to developing negotiation strategies*. Prentice Hall.

Foley, R. M. (2001). Professional development needs of secondary school principals of collaborative-based service delivery models. *High School Journal*, *85*(1), 10–23. doi:10.1353/hsj.2001.0016

Kotter, J. P. (1999).*Leading change*. Cambridge, MA: Harvard Business School Press.

Leithwood, K., Seashore-Louis, K., & Leung, Y. F. (2010). *Conflict management and emotional intelligence*. (DBA thesis). Southern Cross University.

Msila, V. (2011). School management and the struggle for effective schools. *Africa Education Review*, *8*(3), 434–451. doi:10.1080/18146627.2011.618650

Nastase, M. (2007). Types of organisational csonflicts. *Revista de Management Comparat Internaţional*, *8*(4), 80.

Rahim, A. (2001). *Managing conflict in organizations* (3rd ed.). London: Quorum Books.

The Davies Report. (1994). *The 'great battle' in Swansea. Times higher education supplement.* Bristol, UK: Thoemmes Press.

Wanger. (2008). *The global achievement gap: Why even our best schools don't teach the new survival skills our children need---and what we can do about it.* Academic Press.

ADDITIONAL READING

Owens, R. E., & Valesky, T. C. (2014). *Organizational behavior in education: Leadership and school reform* (11th ed.). Allyn & Bacon Educational Leadership.

KEY TERMS AND DEFINITIONS

Complexity Theory: The study of complex and chaotic systems their influence and how order, pattern, and structure can arise from them. Through this theory the role of that is played by large number of seemingly independent agents can spontaneously order themselves into a coherent system.

Conflict: Conflicts are natural and un avoidable phenomenon and behaviors that exist among people through communication process as the internal or external discord that results from differences in ideas, values, or feelings between two or more people.

Conflict Management: Management is the process of reaching organizational mission and goals by working with and through people through utilizing the available organizational resources.

Education: The process of receiving or giving systematic instruction, especially at a school or university. Technically is a process of acquiring knowledge and skills through the process of teaching and learning to change attitude and behavior.

Chapter 14
Leadership Skills in Complex Collegial Adult Groups

Muhamadi Kaweesi
Islamic University in Uganda, Uganda

ABSTRACT

Collegial groups of professional people have been shown to produce high adult achievement and performance and therefore learning the skills of working with groups to solve instructional problems is a critical task of supervision. This paper covers knowledge, skills, and procedures for developing productive instructional improvement groups in the complex school environment characterized by turbulence and chaos emanating from dysfunctional group members. Throughout this paper, the importance of these methods in developing group leadership skills has been explained: using group observations, changing group leadership styles, dealing with dysfunctional members, and preparing for meetings. The paper also demonstrates how development supervision can be applied to professional groups by showing how a developmental supervisor matches the supervisory approaches to the group's characteristics.

INTRODUCTION

Learning the skill of working with groups to solve instructional problems is a critical task of supervision. Collegial adult groups have been shown to produce higher adult achievement and performance than individualistic or competitive learning

DOI: 10.4018/978-1-5225-0460-3.ch014

(Yorks & Kasl, 2002). Professional people who are brought together to deal with pressing mutual problems have a right to expect results. Lengthy meetings break the morale of group members. They might become reluctant a pathetic, and sometimes hostile towards future meetings. They might even suspect that the group leader is deliberately leading them a stray so that the groups' inability to decide can be used as an excuse for the leader to do whatever he or she wishes. Regardless of whether the leader is actually trying to create confusion or truly desires a group decision, the lack of clear results erodes unity and common purpose. Groups that work productively, efficiently, and harmoniously generally have a skillful leader. Unfortunately, since being part of a group is such an everyday occurrence in professional life, we seldom stop to think about what makes some groups to proceed naturally in a professional manner. A leader needs to be conscious of the elements of a successful group, select clear procedures for group decision making, be able to deal with dysfunctional behavior, use conflict to generate helpful information, and determine appropriate leadership style.

Dimensions of an Effective Group

There are two dimensions of an effective professional (Cooper, 2002). The *task* dimension and the *person* dimension. The *task* dimension represents the content and purpose of the group meeting. The task is what is to be accomplished by the end of the meetings. Typical tasks of professional groups might be deciding on a new policy, writing a new instructional schedule, coordinating a particular curriculum, or preparing a professional development plan. An effective group, obviously, accomplishes what it sets out to do. Specific task behaviors are clarifying the group's purpose, keeping discussions focused, setting time limits, and appraising group progress toward the goal. A leader who says, "We're getting off the track; let's get back to discussing real issues," is exhibiting a task behavior. Imagine a group that exhibits only task behaviors. The meeting would be formal, cold, and tense. People would not receive feedback, would not be encouraged, and probably would swallow hard before addressing the unsmiling, staring faces. Such a group would accomplish its task quickly, with little mutual support. The decision would be quick because participants would wish to remove themselves from the tense environment as soon as possible. The formality of the sessions would prevent in-depth discussions of feelings, attitudes, and differences of opinion. Decisions would be made on the basis of incomplete information and commitment from group members. The implementation of the decision would be problematic at best.

The *person* dimension of an effective group comprises the interpersonal process and the satisfaction participants derive from working with each other. Concern and sensitivity to participants' feelings create a climate of yearning to meet with each other from week to week to accomplish and implement the group task. A group that exhibits person behaviors would be characterized by humorous storytelling, and frequent back slapping and touching. People would be smiling and laughing. The image of a wild cock-tail party might characterize a group with all person behaviors and no task behaviors. People would enjoy each other's company for its own sake, everyone would have a wonderful time, but sometimes, little would be done. Specific person behaviors seen in a group include recognizing people for their contributions, smiling, injecting humor, and listening attentively. A leader who says, "Ahmed, I'm following what you've been saying; it's a point worth considering," would be demonstrating a person behavior.

Group Member Roles

First, the leader needs to determine what behaviors are indicative of roles already in existence. Are some members displaying *task roles* and / or *person roles*? What roles are ongoing? Are certain roles lacking. Remember that both task and person roles are functional to group performance. Another set of roles and behaviors, called *dysfunctional*, distract a group from task and person relations. *Dysfunctional roles*, unlike functional roles, are a concern when present. After listing and briefly describing the most common functional member's roles, we will examine dysfunctional roles.

Task Roles

The knowledge of task roles and behaviors provides a guide to the group leader. Consider the following descriptions:

- **Initiator-Contributor:** Proposes original ideas or changed ways of regarding group problem, goal, or procedure. Launches discussion, moves group into new areas of discussion.
- **Information Seeker:** Asks for clarification in terms of factual adequacy. Seeks expert information and relevant facts.
- **Opinion Seeker:** Asks for clarification of values pertinent to the group undertaking or to proposed suggestions. Checks on others' attitudes and feelings toward particular issues.
- **Information Giver:** Provides factual, authoritative information or gives own experience relevant to the issue.
- **Opinion Giver:** Verbalizes his or her own values and opinions on the group problem; emphasizes what the group should do. .
- **Elaborator:** Picks up on others' suggestions and amplifies with examples, pertinent facts, and probable consequences.

- **Coordinator:** Shows the link between ideas and suggestions, attempts to pull diverse proposals together.
- **Orienter:** Clarifies the group's position, gives a state-of-the-scene review. Summarizes what has been discussed, points out where discussion has departed from the goal, and reminds the group of their ultimate goal.
- **Evaluator-Critic:** Evaluates the proposals of the group against a criterion of effectiveness. Assesses whether proposals are "reasonable," manageable," "based on facts," and derived through fair procedures.
- **Energizer:** Focuses the group to move toward decisions. Challenges and prods group into further action.
- **Procedural Technician:** Facilitates group discussion by taking care of logistics. Sees that the group has the necessary materials for the task (paper, pencils, chalk, and so on).
- **Recorder:** Writes down the group's suggestions and decisions. Keeps an ongoing record of what transpires in the group.

A group needs these member roles to keep moving toward accomplishing its task. A leader can use these descriptions to figure out what roles are missing. Additional roles might need to be assigned to group members or incorporated by the leader. For example, if a group has many opinion givers but no information givers, then decisions would be made on the basis of feelings, without regard to actual experience or knowledge. A leader would need to consider ways to add more information giving. Perhaps he or she could assign people to gather more knowledge or ask outside experts for assistance. Likewise, if a group has many opinion givers and information givers but lacks orienters and coordinators, the members may be talking past each other. There would be a lack of direction and a lack of synthesis of the relationships among members' ideas. The leader would need to plan ways to coordinate discussions. As a final example, a group might contain most of the task roles except for a procedural technician or recorder. Such a group probably would converse easily but would bog down on recalling what has been said. The leader who knows what roles are needed can ask for a volunteer to be a recorder and summarizer. Knowledge of task roles and behaviors enables a leader to assess what roles are evident and what further roles need to be assigned. The leader might take on some of the missing roles, assign them to others, or add particular persons to a group.

Person Roles

The knowledge of person roles and behaviors provides a guide to the group leader. Consider the following descriptions:

- **Encourager:** Affirms, supports, and accepts the contribution of other members. Shows warmth and a positive attitude toward others.
- **Harmonizer:** Conciliates differences between individuals. Looks for ways to reduce tension between members through nonthreatening explanations and humor.
- **Compromiser:** Offers to change his or her proposals for the good of the group. Willing to yield position or to acknowledge own errors by meeting other opposing ideas halfway
- **Gatekeeper or Expediter:** Regulates flow of communication by seeing that all members have a chance to talk. Encourages quiet persons to speak and regular puts limits on those who dominate the conversation. Proposes new regulations for discussions when participation becomes unbalanced.
- **Standard Setter – Ego Ideal:** Appeals to group's pride by not letting group members give up when trouble occurs. Exudes confidence that the group is a good one and can make sound decisions.
- **Observer and Commentator:** Monitors the working of the group. Records who speaks to whom, where and when most roadblocks occur and the frequency and length of individual members' participation. Provides feedback when the group wishes to evaluate its procedures and processes.
- **Follower:** Is willing to accept the decisions of the group and follow them even though he or she has not been active or influential in those decisions.

The seven person roles provide human satisfaction and group cohesiveness. People feel positive about meeting and talking with each other and comfortable enough to express their ideas. As a result, meetings are seen as pleasant times to continue the group's work. When person roles are missing, a group may face severe difficulties in making acceptable and committed decisions. Without person behaviors and roles, only the strongest, most assured, and vocal members will speak. Decisions might be made that more timid persons strongly reject but the group may not know that such strong disapproval exists. Again, it is the group leader's responsibility to see if people roles are evident. If roles are missing, then he or she can confront the group with their absence, pick up the role(s) himself or herself, quietly suggest particular roles to existing members, or add to the group other individuals who more naturally play such roles. *Both task and person roles, when not already in existence, need to be added.*

Dysfunctional Roles

Dysfunctional roles and behaviors are those that are conspicuous in their presence. Such roles and behaviors disrupt the progress toward a group goal and weaken group cohesiveness. Consider the following:

- **Aggressor:** Personally attacks the worth of other members. Belittles and deflates the status, wisdom, and motivation of others. Examples of such verbal attacks are: "That's the most ridiculous thing I've ever heard," "You must be crazy to suggest ...."
- **Blocker:** Sees all opinions and suggestions by group members as negative. Opposes any decision being made and stubbornly refuses to propose alternatives. Examples of such blocking statements are: "That's a terrible idea," "I don't want to do that," "It's futile to do anything."
- **Recognition-Seeker:** Uses the group setting to receive personal attention. Examples of such behaviors are dropping books, scattering papers, coughing incessantly, pretending to be asleep, raising hand and then forgetting what one would have said.
- **Self-Confessor:** Uses the group to ventilate personal feelings not related to the group's tasks. Talks about personal problems or feelings of inadequacy whenever he or she can see ways to slip such confessions into the group discussion. Examples of self-confessing statements are: "This discussion reminds me of when I was a little child and the weight problem I had."
- **Playboy or Playgirl:** Lacks interest and involvement by using the group setting to have a merry time. Distracts other members from the group's purpose. Tells private jokes, passes notes, makes faces at others, and so on.
- **Dominator:** Asserts superiority in controlling group discussion and dictates what certain members should do. Claims to know more about the issue under discussion and have better solutions than anyone else. Has elaborate answers to almost every question and monopolizes the discussion.
- **Help-Seeker:** Tries to gain group's sympathy by expressing feelings of inadequacy or personal confusion. Uses such self-derogation as reason for not contributing: "This is all too confusing for me," "I can't make a decision on my own," "Why ask me? I can't help."
- **Special-Interest Pleader:** Has no opinion or suggestions of his or her own but instead speaks for what others would say or do. Cloaks own bias by using an outside group: "We couldn't do that. Do you know what the school board would think?" "If those parents down in the local restaurant ever heard that we were going to change ...."

Dysfunctional roles are fairly self-evident in a group. The leader's responsibility is to reduce or eliminate such dysfunctional roles before they severely harm the morale and efficiency of the group. He or she might try to understand the dysfunctional member's reason for acting as an aggressor, playboy, special interest pleader, and so on, and then might either confront the person privately or provide changes within the group to satisfy the unmet needs that are leading to the dysfunctional behavior. Methods for dealing with dysfunctional behaviors will be discussed shortly, but first let's focus on leadership styles mated with maturity levels of groups.

Changing Group Leadership Styles: A Complexity View

If a group lacks either task or person behaviors, the leader can choose a style that will fill the void. A group that exhibits much initiative, information, and competitiveness (high task) as well as hostility, aggression, and bitterness (low person) could benefit from a leadership style that is encouraging, praising, harmonizing, and humorous (high person). A group that exhibits much positive camaraderie (high person) but is being uninterested, apathetic or uninformed (low task) could benefit from a leadership style that presses for information, sets goals, and enforces procedures (high task). Hersey and Blanchard (1988) in their "*situational leadership*" theory proposed a leadership style that can respond to group characteristics. They identified four styles of leadership based on the relative emphasis on task and relationship (person) behavior:

Style 1 (s1): High task, low relationship. This is an autocratic style, whereby the leader tells the group members what is to be done, when, and by whom. The leader makes decisions for the group. This style is similar to directive supervision. One word that describes this style is *telling*. The leader determines both the process and the content of decision making.

Style 2 (s2): High task, high relationship. This is a democratic style whereby the leader actively participates with the group both as a facilitator of the decision-making process and an equal member contributing his or her own ideas, opinions, and information. This style is similar to collaborative supervision. One word that describes this influencing style is *selling*. The leader attempts to influence both the processes and the content of decision making by being persuasive.

Style 3 (s3): High relationship, low task. This is an encouraging and socializing style whereby the leader promotes cohesion, open expression, and positive feelings among the members but does not influence or interfere with the actual decision. (The leader's role is one of clarification, encouragement, and reflection.) The style is similar to nondirective supervision. Note that the leader participates by helping members express their ideas, opinions, and needs but does not participate in the sense of offering his or her own ideas, opinions, and needs. The leader participates in the process but not in the content of decision making.

Style 4 (54): Low relationship, low task. This is a hands-off or laissez faire style whereby the leader turns the task over to the group and does not participate in any manner. The leader tells the group what the task is and then physically or mentally removes himself or herself from any further involvement. One word that describes this style is *delegating.* The leader is involved in neither the process nor the content of decision making.

Hersey and Blanchard (1988) stated that effective leadership is based on matching leadership style to the readiness of the group. The readiness of a group depends on the particular task; the same group could be of high readiness for one task and low readiness for another. Readiness can be assessed according to the characteristics of ability and willingness. *Ability* is the knowledge, skills, and experience to achieve without the need for outside assistance. *Willingness* is the degree of motivation, confidence, and interest in accomplishing certain tasks. The leader can assess the readiness of individuals and a group according to these levels (Hersey and Blanchard, 1988, pp. 176-177)

- **Readiness Level One (R1):** Unable and unwilling, unable and insecure.
- **Readiness Level Two (R2):** Unable but willing, unable but confident.
- **Readiness level three (R3):** Able but unwilling, able but insecure.
- **Readiness level four (R4):** Able and willing, able and confident.

Situational leadership matches leadership style to the readiness level of the group (see Figure 1 on matching and directionality of a developing group). An R1 group is best matched with a telling autocratic style (S1). An R2 group is best matched with a selling, democratic style (S2). An R3 group is best matched with a participating, encouraging style (S3). An R4 group is best matched with a delegating, laissez faire style (S4)

Groups are complex human entities that respond to the gradual shifting of a group leader's external control in the same manner that an individual teacher will respond to gradual shifting of supervisory control. In other words, an R1 (low readiness) group with an S1 (telling and autocratic) leadership style will not develop until the leader gradually allows them to gain greater internal control. An unmotivated group might work most efficiently with Leadership at first. As the group gains experience, as members become acquainted with each other, and as the group acquires expertise, the leader should be alert to those signals of increasing readiness and provide for greater group involvement by shifting to an s2 leadership

Figure 1. Matching and directionality of a developing group

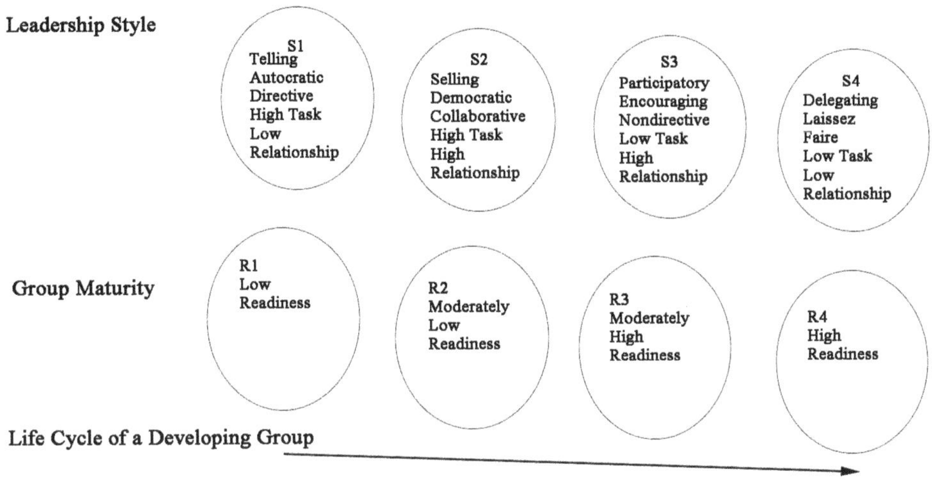

style. It is conceivable that a group working on a long project might complete the entire life cycle by beginning with s1 (telling) leadership and concluding with s4 (delegating) leadership. A group leader might work toward eliminating his or her control over the group. The ultimate goal should be for a group to provide its own task and person behaviors and not be dependent on formal leadership.

Applying Developmental Supervision to Groups

Developmental supervision, when applied to groups, has much in common with situational leadership but also has some key differences. Let us first review how developmental supervision is used with groups, and then compare the developmental model with situational leadership. The developmental supervisor matches one of the four supervisory approaches to the group's characteristics:

- **Directive Control Behaviors:** Are used with a group functioning at very low developmental levels, lacking the expertise needed to act on the problem, and with- out any commitment to its resolution, or in an emergency situation. In group meetings, a supervisor uses directive control behaviors to give a clear message about what changes are expected. The supervisor *presents* by stating his or her under- standing of the problem, *clarifies* by asking group members if they have any perceptions to add, *listens* to input, and then *problem solves* by reassessing the problem and possible solutions. She or he then *directs* by stating what is to be done, *clarifies* by asking for additional input, *standardizes* by laying out a specific time- line and expectations, and *reinforces* by monitoring the expected group performance. The supervisor moves away from directive control and toward directive informational behaviors as soon as possible.

- **Directive Informational Behaviors:** Are used with a group functioning at fairly low developmental levels, with little expertise, and little commitment. The supervisor *presents* by stating his or her understanding of the problem, *clarifies* by asking group members for input, *listens* to their perceptions of the problem, and *problem solves* by formulating two or three alternative solutions. He or she *directs* by stating the alternatives, *listens* after asking the group for input, and *directs* by reformulating and stating final choices. The supervisor *clarifies* by asking the group to choose from alternatives, *standardizes* by suggesting a timeline and criteria for success, and *reinforces* by suggesting ways to follow-up.
- **Collaborative Behaviors:** Are used when the group is functioning at moderate or mixed levels of development, the supervisor and group have the same degree of expertise concerning the problem, and both the supervisor and group are com- mitted to solving the problem. The supervisor *clarifies* by asking group members for their perceptions of the problem, *listens* to group perceptions, and *reflects* by verifying group perceptions. The supervisor *presents* by adding his or her perceptions, *clarifies* by determining if the group understands supervisor perceptions, *problem solves* by exchanging suggested solutions, *encourages* by accepting conflict, and *negotiates* a mutually acceptable solution. Finally, the supervisor *reflects* by summarizing the collaborative plan.
- **Nondirective Behaviors:** Are used when the group is functioning at high developmental levels, possesses extensive expertise, and is highly committed to solving the problem. The supervisor *listens* as group members provide their perceptions of the problem, *reflects* by paraphrasing group members' perceptions and feelings, *probes* the group for additional information on the problem's underlying causes, *listens* further to teacher perceptions, and *reflects* further by continuing to paraphrase group members' comments. The supervisor *problem solves* by asking the group to think of possible actions and to consider the consequences of potential actions, *presents* by asking the group to commit to a plan of action, *standardizes* by asking the group to establish a time line and criteria for success, and *reflects* by restating the group's plan.

The long-term goal of the developmental supervisor is to foster the group toward higher levels of development, expertise, and commitment. This can be accomplished through training in collaborative skills (communication, decision making, problem solving, and so on), providing observer feedback on group behaviors, facilitating self-study, facilitating reflection on group process, and a variety of other strategies for group development. As the group develops, the supervisor gradually shifts from directive to collaborative or from collaborative to nondirective behaviors.

Dealing with Dysfunctional and Chaotic Members

The fact that a group is made up of individuals with varying temperaments and motivations is important when thinking about ways to work with groups. Dealing with individuals, particularly those who display dysfunctional behaviors, is an additional responsibility of a group leader. The leader has observed the group at work and has determined that his or her own leader behaviors are appropriate for most members of the group, yet there continue to be a few dysfunctional members, then individual treatment might be in order. The procedure for treating a dysfunctional member is (1) observe the member, (2) try to understand why the member is acting unproductively, (3) communicate with the member about the behavior, (4) establish some rules for future behavior, and (5) redirect the unfavorable behavior (Corey & Corey, 1982). Each step of this procedure will be amplified.

- **Observe the Member:** When and with whom does the dysfunctional behavior occur? What does the group member do, and how do others respond? For example, a dominator might start monopolizing the conversation as soon as he or she walks into the meeting. Other people might be interested in the dominator's talk for the first few minutes of the meeting but become increasingly annoyed as the dominator continues. They might roll their eyes, yawn, fidget, or make comments to each other.
- **Try to Understand the Member:** Why does the member persist with dysfunctional behavior? Does he or she know the behavior is unproductive? Is the behavior being used to mask some underlying emotion? For example, a playboy might be insecure about his own worth and pretend not to care rather than exposing inner thoughts to the scrutiny of the group.
- **Communicate with the Member:** What can you communicate about the group member's behavior and the situation? Describe the situation and the behavior to the member without maligning. Instead of saying, "You're being an aggressive," say, "I've noticed that you speak loudly and angrily to Aminah. At the last meeting, you told her to keep her mouth shut." Tell the dysfunctional member the effect of the behavior on you as group leader:

"When you tell Aminah to shut up and tell Ahmed that he's stupid, it starts an argument that takes time away from the meeting. I can't complete the agenda on time when those arguments take place."

- **Establish Some Rules for Future Behaviors:** Either ask the member to suggest some rules that he or she can abide by in the future, or tell the member your future rules, or establish them jointly. Regardless of which tactic is chosen, the leader should think of rules that would minimize further disruptions to the group. For example, the leader might say to a self-confessor, "The next time you have a personal problem, come speak to me privately about it," or to a dominator, "I'm going to enforce a two-minute limit on every member's participation," or to a blocker, "If you don't think we're on the right course, tell us your objection once and only once."
- **Redirect the Unfavorable Behavior:** Pick up on the group member's dysfunctional behavior, and try to make it functional. A dominator can be assigned the role of recorder, summarizer, or time keeper. A playboy can be given an opening time for sharing a funny story to relax the group before starting official business.

The five steps outlined here will help the meeting leader understand and deal with individual dysfunctional behavior. The steps are predicated on the leader confronting the dysfunctional member in private. Dysfunctional behaviors that occur infrequently and in isolated situations might simply be ignored. Only incessant behaviors that distract the entire group need to be dealt with via direct confrontation. Confrontation is not easy but is necessary at times for the sake of the group.

Resolving Conflict in Collegial Groups

The key to a productive group is the way ongoing conflict is resolved. Conflicts are particular disagreements that occur between two or more members at a particular time. *Conflict is not necessarily dysfunctional.* In fact, research has shown that successful groups exhibit much conflict. A group can make wise decisions only when there is a wealth of information and ideas to consider. Information and ideas are generated through conflict. To suppress conflict is to limit the group's decision-making capacity. It can stir emotions and creativity. Conflict creates tension and frustration that call for action of a different type than the 'usual peaceful' approaches to group problem solving. Conflict can serve as a mechanism for fine-tuning the group's goal hierarchy. A conflict episode in a group, after its resolution, can help clarify the proper power and authority relationships between different group members.

Therefore, the leader should encourage conflict, not stifle it. Of course, conflict, if not handled correctly, can degenerate into adversarial and harmful relations. It is not conflict that is bad; it is the way the leader deals with it that determines its value.

Hence, the leader has to appreciate that conflict is today a given in groups. Its management is one of the most difficult, yet important job for any group leader. Conflict is recognized as a fact of life, and so the issue is not whether to have it, but how to manage it. However, this acceptance of conflict has not always been the prevailing view. Conflict was once thought to be highly undesirable – something to be stifled at almost any cost. Such was certainly the view of the Classical School of organizational thought. Such classical writers as Henri Fayol and Frederick W. Taylor assumed that conflict was a negative force in the workings of the group and so was to be avoided in favor of harmony.

This approach is most evident in the rules and types of organization structures they suggested as models. They spelt out conditions specifically and in detail to try to prevent conflicts. The classicists, thus, assumed system of organizations, once characterized by rules, procedures and structures that were oriented toward order and harmony as the means to organizational effectiveness. Conflict occurs when there is a disagreement over ideas. The leader should keep the disagreement focused on the ideas rather than on the personalities of the members. The following procedure for handling conflict serves as a ready reference for the group leader:

In the first place, it is helpful to ascertain whether conflict *does* exist between the parties or whether a conflict only *appears* (to the parties) to exist. The criterion is whether the two parties seek goals that are actually incompatible. Frequently, what appears to be a brewing conflict between two parties is, in fact, a misunderstanding. By recognizing the problems of distorted perception as discussed above, it is probable that a misunderstanding can be dealt with through explicit goal setting and improved communication. This often requires training individuals and groups on such skills as group goal setting and prioritizing, as well as in communication skills (for example, active listening, seeking feedback to check receiver's perceptions using multiple channels).

If a conflict does exist, however (that is, the parties do have goals that are mutually incompatible), then it is necessary to select a method of dealing with it as productively as possible from among the many options available. The general principle is that a win-lose approach tends to be the least productive, while a win-win approach–in which both parties win something (though not necessarily equally)–tends to be the most productive. In a conflict situation, it is also important to diagnose the extent to which a party is willing to cooperate with another party. There is a need to determine:

- The level of *cooperativeness,* which is the extent to which one wishes to satisfy the concerns of the other, and
- The level of *assertiveness,* which is the extent to which one wishes to satisfy her or his own concerns.

From the above analysis, five perspectives may be used in conceptualizing conflict and behavior:

1. *Competitive* behavior is the search to satisfy one's own concerns at the expense of others if need be. This involves ensuring that nothing is yielded and every advantage is exploited. It is the classic win-lose view of conflict.
2. *Avoidant* (unassertive-uncooperative) behavior is usually expressed by apathy, withdrawal, and indifference. This does *no*t mean that there is an absence of conflict but that it has been conceptualized as something not to deal with. Hence, the latent conflict remains and may be viewed differently at another time.
3. *Accommodation* (high cooperativeness-low assertiveness) is typified by appeasement: one attends to the other's concerns while neglecting his or her own. This orientation may be associated with a desire to maintain a working relationship even at some sacrifice of one's own interest.
4. *Sharing* orientation (moderate assertiveness-moderate cooperativeness) often leads to compromise (trade-offs).
5. *Collaborative* orientation to conflict (high assertive-high cooperative) leads to efforts to satisfy fully the concerns of both parties through mutual problem solving. The solution to the conflict is a genuine integration of the desires of both sides. The concept is win-win.

Conflict cannot and should not be avoided. Conflict, if encouraged and supported, will enable a group to make better decisions. It is the group leader's handling of conflict that makes the difference. The group should have the feeling that it is all right to disagree and that anyone who does disagree will be able to make his or her full position known.

A Contingency Approach to Conflict Resolution

Contingency approaches to management are predicated upon the concept that diagnosis of the situation is necessary as a basis for action. In dealing with conflict, the contingency view holds that there is not one best way of managing it under all conditions, but that there are optimal ways of managing conflict under certain conditions. An important aspect of conflict management, then, is to consider (1) alternative ways of managing conflict and (2) the kinds of situations in which each of these various alternatives might be expected to be the most effective, not only in dealing with the critical issues but also in doing so in a way that strengthens the organization.

Preparing for Group Meetings

A group can proceed more easily with its task if the leader has made certain preparations. Shelton and Bauer (1994) suggest that pre-meeting planning should involve decisions about whether a meeting is necessary, who should attend the meeting, setting, agenda, and preparation. Some especially important aspects of preparing for a meeting include planning an agenda, establishing ground rules, and writing guided discussion questions.

- **Agendas:** A group has to be clear on its task and purpose. Why are they meeting? What are they to accomplish? Is there to be a product? An agenda distributed several days before the actual meeting will inform members of the reasons for the meeting and what will be accomplished. The agenda need not be elaborate by including a brief explanation and a breakdown of items. Time limits for each item provide members with a sense of priorities as well as the assurance that the leader plans to end on time. Keeping to starting and ending times displays respect for group members' personal schedules.
- **Establishing Ground Rules:** Participants will need to know not only agenda items like the meeting's purpose, place, time, and topics, but also behaviors that are expected of them at the meeting. Ground rules can be established in advance concerning any of the following:
    - Type of participation expected (sharing of information, professional dialogue, choosing from established alternatives, brainstorming, problem solving, conflict resolution, and so on).
    - Roles to be assigned (coordinator, time keeper, information giver, recorder, etc.).
    - Interpersonal expectations (everyone contributes, use active listening, criticize ideas but not people, consider each person's views, use agreed upon conflict management strategies, and so on).
    - Decision-making method (decision by averaging individuals' opinions, majority vote, consensus, etc.).
    - Type of follow-up expected (assigned tasks, continued dialogue, classroom implementation, follow-up meeting, and so on).

- **Guided Discussion:** When meeting with a small group to discuss an issue, it is helpful to have in mind the type of questions to ask. Typically, questions to be asked will shift during a meeting. At the beginning of the meeting, the leader usually spends time clarifying the topic for discussion. During the meeting, the leader uses open-ended questions that allow for seeking, elaborating, and coordinating of ideas, opinions, and information. At the conclusion of the meeting, the leader asks questions that summarize what has been accomplished and what remains to be done.

CONCLUSION

This paper has examined the knowledge and skills needed to help professional groups develop. Particular emphasis was put on the supervisor's role in terms of behaving, confronting dysfunctional members, resolving conflict, and preparing for meetings. The theme of looking at professional groups in a developmental manner should be familiar by now. As a group works together, the leader needs to practice skills that enable the group to become less chaotic, more cohesive, responsible, and autonomous. Eventually the leader would hope to lessen his or her own control and influence so that the group becomes a wise and autonomous body.

REFERENCES

Brandt, R. (1987). On cooperation in schools: A conversation with David and Roger Johnson. *Educational Leadership*, *45*(3), 14–19.

Cooper, S. M. (2002). Classroom choices for enabling peer learning. *Theory into Practice*, *41*(1), 53–57. doi:10.1207/s15430421tip4101_9

Corey, G., & Corey, M. (1982). *Groups: Process and practice*. Monterey, CA: Brooks/Cole.

Fullan, M. (2000). *Change forces: The sequel*. Philadelphia, PA: George H. Buchanan.

Gibson, J.L., Ivanvich, J., Donnelly, J., Jr., & Konopaske, R. (2006). *Organizations, behaviour & structure processes* (12th ed.). McGraw-Hill.

Gordon, T. (1980). *Leadership effectiveness training- E. T.* New York: Bantam Books.

Hare, A. P. (1982). *Creativity in small groups*. Beverly Hills, CA: Sage.

Hersey, P., & Blanchard, K. H. (1988). *Management of organizational behavior: Utilizing human resources* (5th ed.). Englewood Cliffs, NJ: Prentice Hall.

Hersey, P., Blanchard, K. H., & Dewey, E. J. (2002). *Management of organizational behaviour* (8th ed.). Prentice-Hall.

Johnson, D. W., & Johnson, R. T. (1987a). *Joining together: Group theory and group skills* (3rd ed.). Englewood Cliffs, NJ: Prentice Hall.

Johnson, D. W., & Johnson, R. T. (1987b). Research shows the benefits of adult cooperation. *Educational Leadership*, *45*(3), 27–30.

Kasl, E., & Yorks, L. (2002). Collaborative inquiry for adult learning. *New Directions for Adult and Continuing Education*, *94*(94), 3–11. doi:10.1002/ace.54

King, A. (2002). Structuring peer interaction to promote high-level cognitive processing. *Theory into Practice*, *41*(1), 33–39. doi:10.1207/s15430421tip4101_6

Mullins, L. J. (2010). Management & organizational behavior (9th ed.). Pearson Prentice Hall

Shelton, M. M., & Bauer, L. K. (1994). *Secrets of highly effective meetings*. Thousand Oaks, CA: Corwin Press.

Slavin, R. E. (1987). Cooperative learning and the co- operative schools. *Educational Leadership*, *45*(3), 7–13.

Yorks, L., & Kas, E. (2002). Learning from the inquiries: Lessons for using collaborative inquiry as an adult learning strategy. *New Directions for Adult and Continuing Education*, *94*(94), 93–104. doi:10.1002/ace.63

Chapter 15
Decision Making and Problem Solving in Higher Educational Institutions

Muhamadi Kaweesi
Islamic University in Uganda, Uganda

Farooq Miiro
Islamic University in Uganda, Uganda

ABSTRACT

As it is stated that the economy of a country can never be greater than the education of its people, today parents and several organizations are investing a lot of money in education to better the future of their children and also meet the demands of the several generations to come. Through this process, the level of competition among higher education institutions is too high for them to survive and continuously attract parents, citizens and funders to meet demands of both institutional and community challenges. The aim of this paper, therefore, is to explore the concept of decision making, factors that determine the decision making process, the effectiveness of decision making and problem solving in higher education institutions deemed to be largely chaotic and complex. World over, government and states are investing in higher education to transform their communities in terms of knowledge, skills, culture and economic development.

DOI: 10.4018/978-1-5225-0460-3.ch015

INTRODUCTION

As the population grows all over the world, higher education becomes diverse and accessible from every angle in the world to serve society (Craig, 2012), multiple demands and challenges towards institutions of higher education also increase thus provoking decision making to address challenges of real life which seem to be complex (Tucker,1992). For example globalization, higher speed technological advancement and multicultural dimension call for complex decision making and solutions to problem solving to make life meaningful (Jeongmin, 2009). Managers and administrators at the pivot center and steering level of higher education end up seeking different opinions and ways through which the complexity of demands of the world must be addressed (Jonasse, 1999; 2000). The aim of higher education should be making human life better through decision making and problem solving by giving alternative solutions to challenges faced and arrive and conclusive actions (Gagne, 1980). Through this process, complex challenges are tackled with complex influences that form and shape the decision making processes for both academic and nonacademic departments to overcome a myriad of challenges at all levels (Sullivan Syn.).

For the society to progress, higher educational institutions, must play an important role through provision of the necessary human capital for all fields and the only way to achieve this objective, is through stable environment and continuous sustainable academic instruments with sound decision making and systems to influence the curriculum, content, programs and the means through which teaching and learning can be realized.

Its worthy noting that, the major role of HEIs have is to determine the economic level and social transformation of its citizens through education, research and innovation and this should be achieved through the quality of its inputs and outputs in terms of human capital at all levels (Woom, 2005). Therefore higher education services should be comprehensive in nature so that they embed the quality of teaching-learning, enhance research, innovation, and intensify internalization, co-curricular activities to meet global demands (Chew & Tan, 2007). This can only be achieved through complex quality procedures and processes of decision making and problem solving.

To achieve the above mentioned quality through tangible means, HEIs must enhance complex strategic decision making processes through action plan objectives to determine the future of their states through their inputs and outputs (students). To walk the talk, lies in the hands of the top management leadership, their think tanks and the systems employed to achieve these tasks. In the strategic management approach to decision making, action plan and problem solving, HEIs are always advised to conduct an overall study to redefine the direction and strategy of the organization's mission and vision for a better future. While carrying out this task, the past and the current position of the organization should be critically assessed to determine the desired future. (West-Burham, 1994). However there are always obstacles that are complex in nature which hinder the progress of decision making and problem solving (John & Scholes, 1993).

Top management and leadership is always criticized and misunderstood by the academic society about decisions made especially if they are not in favour of their business within the educational institutions (Bergquist, 1992; Cohen & March, 1983). Part of the problem may depend on the sharp brains that institutions employee whereby each feels should contribute or participate in decision making processes and forget that the key responsibility of the top management and leadership should be studying the horizon and move the institutions ahead. This kind of scenario has been experienced all over the world where each staff who attains minimum qualifications as per the national council requirements can be a leader (Bennis & O'toole, 2005). However it has also been established that part of the obstacles caused are the top management persons in that they sometimes tend to neglect consultation of stakeholders when it comes to decisions that pertain to their departments and feel that they should engage them at the implementation level of which in most cases affects the quality of results achieved due to sabotage or selfish reasons (Raush, 2007)

Complex Features in Decision Making

Scholars of educational management and administration have developed theories in different generations with an attempt to understand the mechanics of individual decisions or evaluate their effectiveness in as a single discrete determination action to be taken (Sanfey & Change, 2008), social action to overcome a challenge (Pamerol & Adam, 2002). Roberto (2001) recognizes decisions as processes occurring over time and across cultures, meanwhile Johnson,& Kruse (2009) view it as a choice of commitment against other alternatives, and in some cases an intention leading to logical human action (March, 1991) or even a conclusion to reading and thinking (Critchley,1999). It can be viewed that decisions made with institutions, have different types and models that are helpful which managers and implementers often use to meet daily organizational business and vision (Ikhfan, 2012).

Decision-making is a routine management activity that happens at all levels within an organization with conscious deliberate choices to arrive at a better solution .Inability to make a quality decision may affect every aspect of the organization (Muhammad et al 2009; Al Medlej, 1997). It is intended to execute activities and achieve institutional goals (Ikhfan, 2012).

Decision making in HEIs should aim at strategy for implementation as a key feature in institutional development (Janessen, 2001), through this different units can present their complex challenges, be analyzed in terms of value, integrate knowledge, ideas and skills and then take appropriate steps in problem solving in line with the institutional mission to consider ethical standards. However skills development should be taken seriously in the course of action (Dejong& Van Jooling, 1998). The purpose of decision making and problem solving, should also be helping learners develop higher cognitive skills and this can be achieved through learner-lecturer exposure (Jossen, 2001). Creating such an environment that challenges oriented learning is however complex in nature.

Honoring trust is another important feature in decision making and problem solving that HEIs should protect jealously (Massy, 2003). HEIs should protect and guard the trust and attraction admired by the community. This can be achieved through continuous and sustainable quality decisions made with less expenditure, quality research and protection of values (Massy, 2013). However its complex in nature and it's a big task to achieve as Trow (1994) states "Trust cannot be demanded but freely given". This is brought about by many factors that call for too much attention towards HEIs as it was in the past (Vroeijenstijn, 1995). However presently any quality decision that is taken and achieved in HEIs is always associated with quality of leadership that steers its decision style that later determines that leadership and management style used.

The model of governance and management is another important aspect in HEIs that determines decision making. Since 1980's, the traditional model of management and governance has encountered several criticisms insinuating that its outdated, inefficient, ineffective and not result oriented to address new complex challenges that institutions face (Dill,1995). In management theories they are models recognized and their distinctions in the way educational leaders should handle matters in educational institutions of higher learning, and they are as follows; the classical and behavioral or administrative decision model

Under the classical model, the manager of HEIs acts on reliable complete information with certainty to the environment where as in the administrative model, the manager can act with incomplete information in a risky and unclear environment. In a classical model, the problem is defined with knowledge gathered as required with all possible alternatives and the likely consequences and decisions are optimized whereas the latter is the opposite.

The heuristic approach helps the managers of education institutions to simplify decision making. Under this all stakeholders in decision making rely on readily available information, similar circumstances are compared, current situation, as well as the anchoring and adjustment possibilities. Since making a decision is a process that requires different units and stakeholders to join brains to meet the institutional vision and mission, it requires steps to be followed which in most cases are not simple to be followed due to the fact that they are interdependent. Leaders of higher education institutions should be careful and vigilant with them and their results to avoid chaos within the institutions.

Globally HEIs managers are under pressure if the following are not achieved or realized in the system's performance, accountability, responsiveness, efficiency and effectiveness. At the same time entrepreneurship and self-sustainability should be exhibited (Rosa, Saraiva & Diz, 2005). In the recent past more institutions have mash roomed in the market to compete with state or government owned institutions in terms of quality and self-reliance (Mack, 2003). It should therefore be concluded that if HEIs are to remain relevant in the world market, they must change attitudes and redefine their approaches towards decision making and problem solving.

In the context of Higher Education Institutions, like in any other organization, the execution of decisions is normally done by the top management levels of HEIs. Therefore, the management must have skills in terms of making and taking the decisions. In order to make a quick, accurate and effective decision, it needs to be supported by data and information. Therefore, availability of information plays an important role in the decision making process. It assumes that only with adequate information a decision can be taken accurately. Brinckloe (1977 cited by Salusu, 2000; 65) argued that facts and information can guide the decision maker to make an accurate and effective decision.

As people make decisions, they participate in different types of decision making behavior due to several elements, units and groups of individuals that make an institution, while arriving at a decision as a process, it is worth noting that emphasis should be put on use of appropriate participation of members so that implementation is achieved and arrived at as a whole. As generations change due to environmental demands, society moves further, through this, different skills and competences also change and this is likely to depend on functional competence or relationship skills (Rausch et al., 2001). Morgan (2003), Wheatley (2003) and Schein (1992) point out that organizational success is now more dependent on relationship skills than it has been in the past.

In the recent years, due to globalization, governments are paying attention to the importance of decision making at higher educational levels. Through this councils for quality assurance have been put in place to supervise the input and out of higher institutions, Both academic and policy documents that are made, should address the major challenges in creating equitable access to good-quality education for both national and international levels. In principle, educational provision can be improved through better management practices, and transparency in the use of resources and accountability to all stakeholders (community, parents, students, teachers, etc.). Resolutions taken are supposed to be in line with demands of the councils governing higher education if they are to continue competing favorably. With the nature of higher education institutions, these are areas of sharp brains where all members of management committees feel that their contributions should be heard, recognized and felt by the systems through this process, sometimes decisions take long to be reached.

It's worth noting that, the ways through which the decisions are arrived at and taken into action, its impact will be felt in the same manner. Therefore managers and administrators of higher institutions have a great task to study the horizon and arrive at decisions that fit all the stakeholders' behaviors and understanding to reach at a common goal. When this process is not followed well in its complexity, mistakes may occur at any stage of implementation and at the end of it all managers may experience problems hence impeding the progress of institutional development and mission. Under a wide range of conditions as stated by Timberlake (2004) increased participation leads to greater 'buy-in,' commitment to decisions, and motivation. Traditional models and understanding of decision-making place too much confidence in the ability of executives to analyze an organization and its environment, that's why it is stated that strategic decision- making should not be isolated at the top of an organization. Distributing strategic decision-making allows people at all levels to make appropriate decisions aligned with one another (Worley, Hitchin &.Ross, 1996).

As there is a broad agreement on participation in decision making, there are also perceived problems including reduced decision-making speed (Daft &. Marcic, 2001) and the dangers of group thinking (Qariis, 1972). Clearly, leaders who are at the pivotal centre of all activities in an institution should be instrumental in maintaining or changing the decisions that do not meet cultures of their organizations (Schein, 1992 and Massarik and Pei-Carpenter, 2002). In the end, situational leadership becomes a prerequisite of an effective decision implementation because a good leader will adapt his or her style to the situation or challenges at hand, regardless of style preference as per the models of decision making (Rowe &L Mason, 1987)

Decision making is a critical aspect to feeling successful and happy in life; decision making is at the root of all we do. It is important to develop effective decision making skills and strategies. Problem solving strategies include, but are not limited to brain storming, cost benefit analysis, written remediation plans, and an examination of possible choices (Wester, Christianson, Fouad, & Santiago-Rivera, 2008). The decision making process can be complicated and overwhelming. As a result, it is valuable for individuals to learn a model to follow, that may be applied to everyday decisions, as well as life changing choices.

Krantz and Kunreuther (2007) posited that a goal and plan based decision making model is an effective and sound approach to take in decision making; in this model, the individual is encouraged to focus on goals, not happiness or usefulness. According to Krantz and Kunreuther, plans are designed to meet one or more goals. That is, people make plans to unconsciously or consciously meet the goals they have. And, some plans satisfy several goals. For example, people who attend a sporting event with a friend may be satisfying several goals; friendship and camaraderie, emotional stimulation from competitive sport, and potentially useful social knowledge gained from watching the game. In this model, goals are context dependent and plans are based on their ability to meet the goals. Essentially, in the goal/plan based model, the context provides the backdrop for the decision that needs to be made; goals and resources, influenced by the context, contribute to the development of plausible plans; while the decision making rules are implemented and influence the plan that is ultimately chosen. Krantz and Kunreuther apply this theory to the insurance business, but imply the theory may be appropriately applied to a variety of contexts.

Factors Leading to Complexity of the Decision Making Process

In the decision making process there are several factors that could influence the decision depending on several studies so far conducted these factors, including past experience (Juliusson, Karlsson, & Gärling, 2005), cognitive biases (Stanovich & West, 2008), age and individual differences (Bruin, Parker, & Fischoff, 2007), belief in personal relevance (Acevedo, & Krueger, 2004), and an escalation of commitment, influence what choices people make. The individual and the organization are also the two "nature" factors influencing the decision making process (Haris, 2012). Blackmore and Berardi (2006) argue that there are at least seven factors, which can influence decision as stated below, decision makers, the environment or condition, the mind to think about the problem or opportunity, the criteria involved, the stakeholders involved and how they are affected, the time as well as decision support theories, models, tools, strategy and technicalities.

Atmosudirjo (1982) described two important factors that influence the process of decision-making: nature of organization and personal capabilities of the decision-maker. The individual or personal factor is considered as the most difficult to control or to predict in the decision-making process, because there are many variables might be involved in this factor (Haris 2012). The nature of leadership in both private and public is seen as another factor that contributes to decision making, where by all major solutions to problem solving depend on the heart of its leadership though sometimes there can be decentralization to only concerned departments within the institution (Elsass, 1997). However for a conclusive action to be implemented, three levels are maintained that lead to proper choice in problem solving, at a personal level an individual goes through the entire process to address the problem depending on its complexity of decisions needed and the resources available, through this course of action, structured and un structured procedures are realized (Fosketta & Hemsley-Broom, 2001).

At the end of the whole process, different ideas from all levels up to the strategic level are concerned and involved till implementation. Arroba (1998) mentioned five factors affecting the decision making process, which are related to the decision-maker (person), namely:

1. Information that was known concerning the concrete problem, which needs to be solved,
2. The level of education,
3. Personality,
4. Coping, in this context can be experience related to the problem (the adaptation), and
5. Culture.

With the global trends, governments and states have also penetrated the way decisions are taken at HEIs. That way state officials feel that their agendas should also be addressed while pushing institutional vision and mission ahead. This is now all over the world in most of the funded and non-funded institutions and failure to comply has somehow caused some institutions disastrous effects or loss of government support and funding.

The style of leadership is another factor in modern situations; leaders must attract the consensus of subordinates other than blind following. To gain loyalty from fellow workers, the top manager should address motivational factors that determine staff performance for example reward, safety needs, promotion, fringe benefits and glory. Emphasis should be put on subtle factors that determine individual participation and implementation of decision (Bmjcareers, 2005). For the decision to be respected and implemented depends on the level at which top leadership involve its team in decision making (Daft, 1997).

Leaders of HEIs should address the changing trends in decision making rapidly to avoid a frustrating circumstance and environment to its customers and other stakeholders (Seymour, 1992). It can further be noted that, the individual or personal factor mostly corresponds to the psychological aspect of the decision-maker, whereas the organizational factor deals more with environment or condition in the organization. Furthermore, organizational behaviour and dynamics are multi-determined and relatively complex. Thus, it calls for examining and understanding the situation in an organization (Cremona 2012).

Syamsi (2000) identified four factors that influence the process of decision-making. These factors are:

1. State of organization,
2. Availability of information,
3. External condition/ environment, and
4. Personality and skill of decision maker.

Diversity and the value of an organization attached to it is another important factor. So far the compiled studies since 1980's to date demonstrate the advantage that teams with vast knowledge and experience have over others (Johnson & Packer, 1987, May & Whitney, 1995). Such teams are advantaged when it comes to implementation and vast experience out sourced from different levels of education and experience (Watson, Kaumar & Michealson, 1993)

The dynamic nature of an organization is also considered as another instrumental factor, which can influence decision making process. Siagian (1987) explained the dynamics in the organization in three categories:

1. The dynamics of individuals within the organization,
2. Group dynamics in the organization, and
3. Dynamics of the organizational environment.

The approach and step in the implementation of the decision should be materialized through good leadership from management. Through the right approach, the people, who have to implement the decision or who will be affected, could carry out the decision that was given to them well and happily, without force and pressure. According to Copeland (1998, cited by Syamsi, 2000), after a decision is made, there are complexes that staff may experience in higher education as a result of a variety of reactions. In addition, present decisions influence future decision making. Several of the outcomes that may result from a decision are regret or satisfaction; both of which influence upcoming decisions.

Regret, feelings of disappointment or dissatisfaction with a choice made is one potential outcome of decision making. Interestingly, regret may shape the decision making process. According to Abraham and Sheeran (2003), anticipated regret is the belief that the decision will result into an action. Anticipated regret may prompt behavior; that is, when a person indicates they will do something, such as exercise, they may follow through with their intended decision, to avoid regret. Once the decision is made, the impact of the decision, if regret is experienced, will impact future decisions. People can often get consumed with examining the other options that were available; the path not taken (Sagi & Friedland, 2007). People feel regret in accordance with how the decision was made as mentioned by Agi and Friedland (2007) in their theory regret may be dependent on the number of options that were available during the decision making process; and how varied the options were may impact how regret is experienced after the decision was made. Through a series of experiments, Sagi and Friedland concluded that people feel remorse because they feel they were able to make a better choice by looking at more information, previously disregarded, and carefully weighing the pros and cons of each choice.

As explained before, future decision making is based on past decisions, as well as levels of satisfaction or regret (Abraham & Sheeran, 2003; Juliusson, Karlsson, & Garling, 2005; Sagi & Friedland, 2007). Even though there is evidence to support this notion, in many cases, particularly when the decision may be reversed, decisions may be based on the reversibility factor (Gilbert, & Ebert, 2002). Significant to individuals' satisfaction is that people are willing to pay a premium for the opportunity to change their minds at a later date (Wood, 2001). They concluded that people do prefer to have the option to change their minds; although people's ability to change their minds actually inhibits their ability to be satisfied with their choice.

Problem Solving

Problem solving is the way of dealing with any challenge in terms of performance deficiency or an opportunity to advance institutional goals and objectives, it requires a proper process of identifying the discrepancy between the actual and desired state of behavior and take appropriate action. The entire process of dealing with a problem is dependent upon the right people with proper and sufficient information in the right time. Upon this institutional leadership requires to have a team to act as a think tank with sharp brains to advise it accordingly and arrest situations that can cause anarchy in the management of affairs.

This can be backed up with scholarly definitions: in view of the above, Duncker (1945) defined a problem as a situation when a living creature has a goal but does not know how this goal is to be reached, when one cannot go from the given to the desired situation simply by action. Newell and Simon (1972) defined a problem as some blockage in a gap that prevents a person from immediately seeing a course of action and if there is no blockage, then the situation is an exercise, not a problem. Alıcıgüzel (1979) defined problems as the difficulties faced by individuals and communities that need to be resolved in order to achieve success. Ritz, Deal, Hadley, Jacobs, Kildruff and Skena (1986) defined a problem as a need, which must be met. Türer (1992) was of the opinion that if there is no any purpose then there is no problem, in other words, the desire to fulfill a need to achieve a purpose and the difficulties objecting these are the main conditions of a problem. For Erden and Akman (1998), a problem is a new trouble faced by the individual that the individual does not know how to surmount. Woods (2000) viewed problems to be challenges we focus on to solve where there is no immediately apparent procedure idea, or routine to follow.

From the foregoing, a problem has been defined to mean so many different issues to so many different individuals and organizations. However, the fact remains that a problem is still a problem as long as there exists, a reason or a need for improvement. That is the gap between the current situation and a desired situation. Attempt(s) at bringing about the desired is referred to as problem-solving (Mathew, 2014). To different scholars, problem-solving like problem is viewed differently. According to Stones (1994), problem solving includes integration of concepts and skills to get over the unusual complete situations. Krulik and Rudnick (1996) viewed problem solving as one of the primary skills that students must take with them when they leave the classrooms and enter the real world. In the views of Candelaria and Limjap (2002), the development of critical thinking skills essential for problem solving does not necessarily require direct instruction. Students may acquire the skill as they interact with their environment in the school and at home, thus honing their creative skills as well. In a later publication, Mayer and Wittrock (2006) define problem solving as cognitive processing directed at achieving a goal when no solution method is obvious to the problem solver.

It can further be noted that problems are of different types and may require different approaches to resolve them as outlined below; Structured, unstructured and crisis therefore in the recent transition to the information age HEIs has focused attention on the processes of problem solving, management, decision-making and continuous improvement to avoid firefighting (Boland & Collopy, 2004; Stice, 1987; Nickerson, Perkins & Smith, 1985; Whimbey & Lochhead, 1982).

It can further be observed that when institutions become vast, their leaders feel should expand more and more by creating campuses in different regions of the world, as this is implemented, problems occur at campuses still. In these uncertain times, campus stakeholders must learn to make the best decision possible in ambiguous situations where there is less than optimal information available and the outcome cannot be predicted. Campus leaders can engage their constituents in substantive discussions about the impact of the issues raised by these case studies on their campus, helping faculty, trustees, and administration develop consensus about priorities for engagement, levels of resources required, and/or the need for reallocation of resources.

The cases also can be used in a teaching or training environment. Because these cases outline contemporary campus issues, they are useful in either a higher education administration degree program or a leadership training program managed by a campus or higher education association. In line with the above, Womack and Podemski (1985) have reviewed state supported multi-campus systems and outline the importance of equal participation in planning processes from each campus in a multi-campus system. They further state that each campus in a system should base mission and goals, at least to some degree, on the needs of the individual communities served. Finally, they recommended that different campuses excel in ways that fit their particular environments. Vroom (2000) writes: under a wide range of conditions, increasing participation leads to greater 'buy-in,' commitment to decisions, and motivation to implement them effectively (p. 85).

All problems are not equally important and their relative significance should be kept in perspective. Despite the importance of problem solving, many educational analysts and industry representatives report that students leave higher education with an underdeveloped ability to solve open-ended problems (Commission on Accountability in Higher Education, CAHE, 2005). In part, this arises because instructors of undergraduate courses prefer students to construct knowledge through single-answer analytical problem solving before they address more complicated open-ended problems that require higher levels of knowledge. Where analytical problem-solving tends to invoke cognitive skills primarily, open-ended problem-solving involves significant social and affective dimensions.

As students are part and parcel of HEIs, and they are prepared to deal with society challenges of the changing world, lecturers should think of good problem solving skills that empower students in their educational, professional, and personal lives. Nationally and internationally, there is growing recognition that if education is to produce skilled thinkers and innovators in a fast-changing global economy, then problem solving skills are more important than ever. The ability to solve problems in a range of learning contexts is essential for the development of knowledge, understanding and performance. Requiring students to engage with complex, authentic problem solving encourages them to use content knowledge in innovative and creative ways and promotes deep understanding (Crebert, Patrick, Cragnolini, Smith, Worsfold & Webb, 2011).

It should be further noted that as human beings continue to live as a community with different cultures of different education background, with different capacity in thinking and approaches towards world affairs, problem solving is now a culture within a common place for example in the industries in the process of profiteering, customer satisfaction, safety, security, etc. In the schools, there is often a misconception between problem and exercise. In the real world, exercise and problem differ in many diverse ways therefore managers should use appropriate ways to address issues at hand.

It is worth noting that HEIs prepare students for future generations and many of them recruit their best performers' academic wise to train them further and later take on the mantle form the seniors as time changes. Mourtos, DeJong and Rhee (2004) were of the opinion that "while the exercises make an important first step in helping students bridge the gap between theory and application, they do not provide the depth and complexity necessary to master problem-solving skills... Students who train mostly in exercise solving tends to develop serious handicap. They rely heavily on solutions they have seen before, rather than working from first principles. Thus a problem with brand new context presents a formidable challenge to them." Therefore, problem solving involves error and uncertainty and even if your students are eventually successful, it is likely they will feel uncomfortable, as they come to terms with the problem solving processes they will encounter in the workplace (Ryan, 1996). Whatever form the problem takes and whatever approach is used to help students develop their problem solving skills, it is important to recognize and make students aware of the differences between solving exercises and solving problems.

CONCLUSION

While there is broad support for participation in decision-making, there are also perceived problems including reduced decision-making speed (Daft & Marcic, 2001). Leaders should be very careful and vigilant in the way decision are taken implemented to address the millennium goals both national and international with a foresighted eye that does not jeopardize the efforts of institutional team members, the stakeholders commitment and used all mechanism that can attract support from staff and students to avoid chaos and peculiar situations which can cause bad results on the side of the administrators, the institutions and the community at large

REFERENCES

Abraham, C., & Sheeran, P. (2003). Acting on intentions: The role of anticipated regret. *The British Journal of Social Psychology*, *42*(4), 495–511. doi:10.1348/014466603322595248 PMID:14715114

Acevedo, M., & Krueger, J. I. (2004). Two egocentric source of the decision to vote: The voter's illusion and the belief in personal relevance. *Political Psychology*, *25*(1), 115–134. doi:10.1111/j.1467-9221.2004.00359.x

Alıcıgüzel, İ. (1979). *Education in first and middle degree schools*. Istanbul: İnkılap ve Aka Bookstore.

Arroba, T. (1998). Decision making by Chinese – *US. The Journal of Social Psychology*, *38*, 102–116.

Atmosudirdjo, P. (1987). *Beberapa pandangan umum tentang pengambilan keputusan*. Jakarta: Ghalia Indonesia.

Berardi, A., & Blackmore, C. (2006). *Introducing environmental decision making.* Open University Worldwide.

Brent, D. R., & Susan, J. (2012). *Leading in tough times: Case studies for higher education leaders.* NACUBO.

Bruine de Bruin, W., Parker, A. M., & Fischhoff, B. (2007). Individual differences in adult decision making competence. *Journal of Personality and Social Psychology, 92*(5), 938–956. doi:10.1037/0022-3514.92.5.938 PMID:17484614

Candelaria, M. L., & Limjap, A. A. (2002). Problem solving heuristics of college freshmen: A case analysis. *Asia Pacific Education Review, 11*(2), 199–223.

Cremona, K. (2012). *A framework for understanding organisational dynamics.* Australian Businesswomen's Network. Available from: http://www.abn.org.au/site/article/organizational-dynamics

Daft, R., & Marcic, L. D. (2001). Understanding management. San Diego, CA: Harcourt College Publisher.

Duncker, K. (1945). On problem solving. *Psychological Monographs, 58*(5).

Erden, M., & Akman, Y. (1998). *Education psychology.* Ankara: Arkadaş Press.

Gilbert, D. T., & Ebert, J. E. J. (2002). Decisions and revisions: The affective forecasting of changeable outcomes. *Journal of Personality and Social Psychology, 82*(4), 503–514. doi:10.1037/0022-3514.82.4.503 PMID:11999920

Jullisson, EA, Karlsson, N, Gärling, T. (2005). Weighing the past and the future in decision making. *European Journal of Cognitive Psychology, 17*(4), 561-575. DOI: 10.1080/09541440440000159

Krantz, D. H., & Kunreuther, H. C. (2007). Goals and plans in decision making. *Judgment and Decision Making, 2*(3), 137–168.

Massarik, F., & Pei-Carpenter, M. (2002). *Organization development and consulting.* San Francisco: Jossey- Bass/Pfeiffer.

Mathew, M. (2014). Introducing effective problem solving culture in higher education institutions in Nigeria: Methods and tools. *Journal of Education and Practice, 5*(25).

Mayer, R. E., & Wittrock, R. C. (2006). Problem solving. In P. A. Alexander & P. H. Winne (Eds.), *Handbook of educational psychology* (2nd ed.; pp. 287–304). Mahwah, NJ: Erlbaum.

Mohamed, A., & Abdelraheem. (2005). *Problem solving and decision making published by center of advancement of postgraduate studies and research in engineering sciences*. Faculty of Engineering, Cairo University (CAPSCU).

Morgan, H. (2003). Challenges of a global marketplace. In Human Resources in the 21st Century. San Francisco, CA: John Wiley &. Sons.

Newell, A., & Simon, H. A. (1972). *Human problem solving*. Englewood Cliffs, NJ: Prentice-Hall.

Rausch, E., Halfhill, S. M., Sherman, H., & Washbush, J. B. (2001). Practical leadership-in-management education for effective strategies in a rapidly changing world. *Journal of Management Development, 20*(3), 245–257. doi:10.1108/02621710110386381

Ritz, J. M., Deal, W. F., Hadley, F., Jacobs, J. A., Kildruff, T. F., & Skena, K. G. (1986a). Problem-solving. *Technology Teacher, 46*(2), 15–22.

Sagi, A., & Friedland, N. (2007). The cost of richness: The effect of the size and diversity of decision sets on post-decision regret. *Journal of Personality and Social Psychology, 93*(4), 515–524. doi:10.1037/0022-3514.93.4.515 PMID:17892329

Salusu, J. (2000). Pengambilan keputusan stratejikuntuk organisasi publik dan organisasi non profit. Jakarta: Gramedia Widiasarana Indonesia.

Schein, E. H. (1992). *Organisational culture and leadership* (2nd ed.). San Francisco: Jossey-Bass.

Siagian, S. P. (1987). *Filsafat Administrasi*. Jakarta: PT Gunung Agung.

Syamsi, I. (2000). *Pengambilan keputusan dan sistem ınformasi*. Jakarta: Bumi Aksara.

Timberlake, G. R. (2004). Decision-making in multi-campus higher education institutions. *Community College Enterprise, 10*(2), 91-99.

Wester, S. R., Christianson, H. F., Fouad, N. A., & Santiago-Rivera, A. L. (2008). Information processing as problem solving: A collaborative approach to dealing with students exhibiting insufficient competence. *Training and Education in Professional Psychology, 2*(4), 193–201. doi:10.1037/1931-3918.2.4.193

Wheatley, M. (2003). When change is out of our control. In Human Resources in the 21st Century. San Francisco: John Wiley & Sons.

Wood, S. L. (2001). Remote purchase environments: The influence of return policy leniency on two-stage decision process. *JMR, Journal of Marketing Research, 38*(2), 157–169. doi:10.1509/jmkr.38.2.157.18847

Woods, D. (1985). Problem-based learning and problem solving. In D. Boud (Ed.), *Problem-based Learning in Education for the Professions*. Kensington, Australia: Higher Education Research and Development Society of Australasia.

Woods, D. R. (2000). An evidence-based strategy for problem solving. *The Journal of Engineering Education*, *89*(4), 443–459. doi:10.1002/j.2168-9830.2000.tb00551.x

Worley, C. G., Hitchin, D. E., & Ross, W. L. (1996). *Integrated strategic change*. Reading, MA: Addison Wesley.

Chapter 16
Self-Assessment of Principals Based on Leadership in Complexity

Leyla Yılmaz Fındık
Hacettepe University, Turkey

ABSTRACT

This study is designed to provide a picture of principal leaders in the school. The focus of this study is to identify and describe the characteristics of principal's leadership process in complex situation based on self-assessment. The study conducted by qualitative approach consisted of volunteer principals working in the city center of Ankara, Turkey. We discuss how these findings provide a deeper understanding of the latest principals' leadership process in complexity based on principals' self-assessment. The researcher gathered and analyzed data in an effort to create a template of essential leadership characteristics that could be utilized at other schools or in other educational environment. The results likely to suggest that self-assessment of principals in complexity are well worth revealing how the principals perceive their leadership process and what leading characteristics are viewed when they have complex situation.

INTRODUCTION

Education nowadays is in a mode of change, these changes result in expecting much not only from education but also from educational leaders and moreover the paradigm for school leadership is undergoing a profound change, this result in significantly

DOI: 10.4018/978-1-5225-0460-3.ch016

raised expectations for school leaders. Therefore school leaders face rapidly changing conditions both in the external environment and within the educational leadership field. These changing conditions pose formidable challenges for educational leaders (Hannah, Balthazard, Waldman, Jennings & Tatcher, 2013; McCarthy, 2002; Tobin, 2014). This leads educational leadership to challenge to engage in a broader conceptualization of its purpose (Tobin, 2014; Tillman, 2002). In addition to their traditional role as school managers, administrative leaders are increasingly expected to help teachers adopt risky new instructional strategies, to engage stakeholders in change through participatory decision making, and to collect and analyze data as they lead their constituents through the comprehensive school improvement process (Henning & Robinson, 2002).

Stakeholders from every field of study have concluded that leadership is defined as a central variable in the organizational success (Murphy, Elliott, Goldring, & Porter, 2007). Especially for education, there is parallel evidence that leadership is considered as the keystone element in school to define student achievement (Murphy & Hallinger, 1988; Leithwood, Louis, Anderson & Wahlstrom, 2004; Marzano, Waters & McNulty, 2005). Over the last three decades it is an assertion that not a particular leadership is equal but all types of leadership is especially visible in high performing schools (Murphy, Elliot, Goldring & Andrew, 2007).

Many researches appear to confirm that principal leadership can have an essential but largely indirect effect on student learning or student success (Blandford, 2013; Hallinger, 2010; Marzano, Waters, & McNulty, 2005; Klar & Brewer, 2013; Raja & Palanichamy, 2011; Robinson, Lloyd, & Rowe, 2008; Wahlstrom, Louis, Leithwood, & Anderson, 2010; Ward, 2013). Principals have an impact on student learning by creating conditions at school that would have a positive impact on teacher practice and students learning (Hallinger, 2010; Wahlstrom, Louis, Leithwood, & Anderson, 2010). Still many researches on leadership, school leadership and/or successful school leadership have been being done to expand these earlier researches (Blandford, 2013; Hallinger, 2010; Wahlstrom, Louis, Leithwood, & Anderson, 2010). For these reason much efforts have given to study the impact of principal leadership and to identify professional leadership dimensions of the principal's role that had an impact on school success (Hallinger, 2010, p.61).

It is an undeniable fact that there is a correlation between the school leaders and students success (Marzano, Waters & McNulty, 2005). There are many researches providing lists of the practices and characteristics of effective school leaders. Marzano, Waters, & McNulty (2005), developed a list of 21 categories of behaviors (responsibilities) of school leaders, such as *knowledge of current curriculum, instructional, and assessment practices* in their meta-analysis of over 300 studies regarding school leadership as practiced by principals. All these behaviors were found to be positively correlated to student achievement. Since the last decade there has been an increased interest in raising a global interest in educational reform centering on student learning led, to a focus on school leadership in general (Hallinger, 2010, p.61). This results in exploring the sources, means and implications of viewing school leadership more broadly than that which is exercised by the principal (Gronn 2002; Harringer, 2010; Leithwood, Mascall & Strauss, 2009; Ogawa & Bossert 1995).

Principal as the school leader in traditional structure is seen as knowing everything and is the only person who takes the decisions for the school. This kind of leadership style no longer meets the needs of changing complex school. School leaders are expected to have the capability to understand the need to change and they need to embrace unpredicted situation in educational environment. This makes complexity leadership worth exploring in educational leadership as complexity leadership offers much for school leadership (Morrison, 2015). Complexity leadership focuses on complex system and processes comprising leadership (Uhl-Bien, Marion & McKelvey, 2007).

Underpinning this approach, this study is an attempt at making self-assessment of principals based on leadership in complexity. The shift to school based-management seems to require educational leaders having knowledge, skills and attitudes that are different from those of the past to manage unpredicted complex situation. Thus, to learn the answer of the question "what the knowledge, skills and attitude needed for educational leadership in today's complex schools are" is seems to be valuable from the practitioners working at schools as principals. Knowing that leadership occurs at all levels of management, whether principals are aware of the role of leadership in complexity and how they assess their own leadership activities and what leadership skills they have are really worth studying. Self-assessment of principals can be a fundamental force in achieving success at schools and the reviews will establish what the key elements are in relation to effective leadership, what principles and processes are associated with effective leadership, and what the implications are for school success as a means of leading complexity. The research questions are:

1. What is necessary to be a leader?
2. What is important to be a leader?
3. What make it easy to be a leader?
4. Why do you think you do not have difficulties in leading your colleagues?
5. What are your main problems your have experienced during your leadership at school?

METHODOLOGY

This study adopts a qualitative research approach. Data was collected through a semi-structured interviews consisting of open-ended questions and demographic data such as level of education, years of experience, types of school and title. Experts' opinions were asked to ensure the content validity of the data collection tool in order to produce valid and reliable results. 3 lecturers also had looked at the questions to ensure content validity and then a final interview protocol was produced after reviewing the provided feedback. Two principals, who did not participated in the study group, read the protocol to check whether the semi-structured interview protocol was clear enough for the participants.

Data was collected in the education year of 2014 through face to face interviews with principals. İnterviews were conducted at the room of principals at the school. The interviews took 15-20 minutes. During the interviews, there was no other person in the room. The principals were asked about their work experiences, leadership practices and also what they had learned from their experiences. Examples of the questions asked are the following: What is/are necessary to be a leader, What is/are important to be a leader, why do not you have difficulties in leading your colleagues, what makes it easy to be a leader, what are your difficulties you have during your leadership?

Descriptive analyses were employed for data analysis. After reading several times irrelevant parts were removed and the statements relevant to the study were listed to answer the sub-questions of the study. Reported limitations are based on the study group, comprised of principals working at in the city center, in Ankara.

Study group consisted of randomly selected 20 volunteer principals working at the state school in the city center in Ankara, Turkey. Figure 1 shows educational background of the principals.

Figure 1. Education level of study group

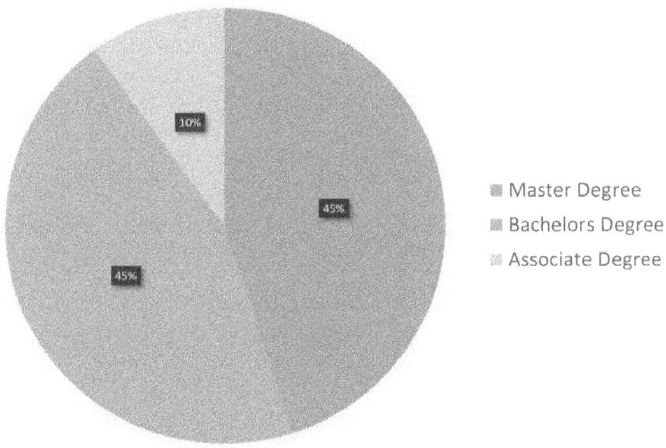

Two of the principals had Associate degree; nine principals had Bachelor's degree and nine of them had Master degree.

Experience year of study group is given in Figure 2.

Six principals had experiences between 0-9; ten of the participants had experience between 10-19 and four of them had 20 or above years of experience.

The share of principals and vice principals are given in Figure 3.

6 of the participants work as a vice-principals and 14 of them work as principals.

Figure 2. Experience year of study group

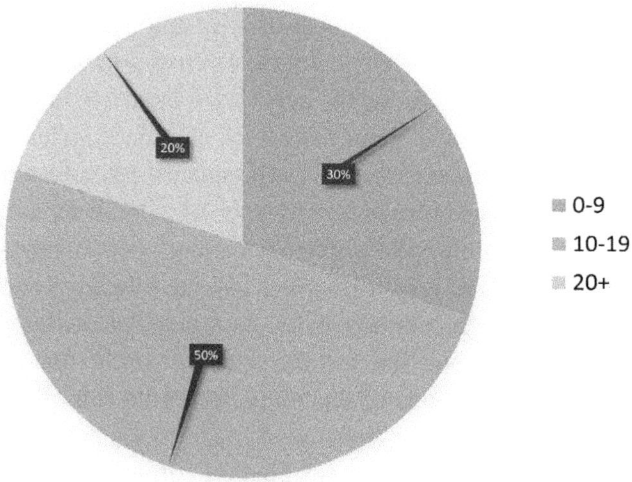

Figure 3. Title of study group

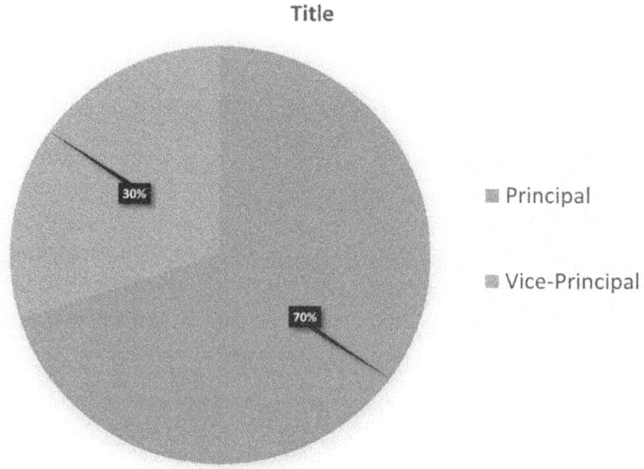

Figure 4. School type of study group

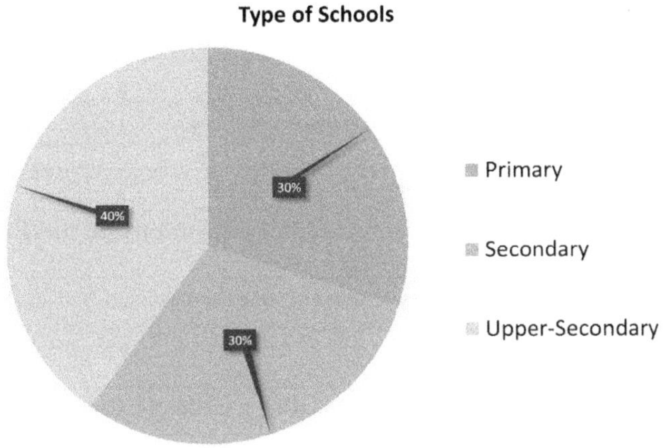

The principals are selected from different types of school; these are primary, secondary and upper-secondary schools. The school types which principals work at are given in Figure 4.

6 of the principals work at primary school; 6 of the principals work at secondary schools and 8 of them at upper-secondary schools.

FINDINGS

The research aims to see how principals assess their leadership process. To learn this, some questions are asked to the principals working at different types of schools. Findings will be presented and analyzed in the order of research questions via tables. Necessary competence to be a leader is given in Table 1.

As shown in Table 1, most principals mention that innovations, vision, being master at work, taking dare and quick decision, comprehending others are really necessary. It is shown that principals in this study also made focus on vision as a keystone element in leadership that stated so many times in many research (Erçetin, 2000; McEwan, 2003; Murphy, Elliot, Goldring and Andrew, 2007; Murphy, Elliott, Goldring, & Porter, 2006; Tobin, 2014; Weng & Tang, 2014). Ward (2013) revealed that a shared vision and goals are one core component that became the school's cornerstone and driving force. It is understood that communication and comprehending each other are also emphasized by the principals as an essential characteristics, this is parallel with the research of McEwan (2003) and Ward (2013). Decision making is emphasized by the principals as an important issue as focused in other research (Gomez, 2013; Murphy, Elliott, Goldring, & Porter, 2006).

Table 1. Necessary competence to be a leader

School Principals	What Is/Are Necessary to Be a Leader?
P1	Innovation
P2	Knowing your colleagues well and making your colleagues be involved in decision-making process
P3	Speaking, listening, understanding and making difference
P4	having mission and vision, having team spirit, quick thinking, taking quick decision, acting with the spirit of cooperation
P5	Having experience, master of his/her job, knowing the regulations and law well, openness to innovations, leading, being honest and protective, increasing confidence
P6	Empathy
P7	Being determined, having the ability of thinking scientifically, having international ethics, being respectful to human right and fair
P8	Leading, setting goals, directing colleagues with his/her professional qualities
P9	Fully comprehending the human relationships
P10	communicating effectively, trusting people and being reliable
P11	Being well-educated
P12	Having empathy, accepting criticism, questioning, being patient, determined, having principles and keeping up with the innovations
P13	Comprehending the colleagues
P14	--
P15	Cooperation
P16	Taking daring decisions and implementing them, the art of speaking in public eloquently and effectively, being charismatic
P17	Finding rational solutions, making colleagues be involved in decision-making process, guiding
P18	Having different skills in different field, providing colleagues with opportunities for professional developments, flexible, openness to innovations, sharing with stakeholders
P19	Empathy
P20	Career

Important competence to be a leader based on the expressions of the principal working at state school in Turkey is given in Table 2.

Table 2 indicates that principals find communication, being determined, having vision and mission, self-confidence, the ability of problem solving and respect are pivotal. Problem solving, consensus building, communication are indicated as effective skills in the study of Marzano, Waters & McNulty (2005) and Murphy, Elliott, Goldring, & Porter (2006). Being honest, empathy, experience and objectivity are stated as the more important features by principals. Motivating other, making

Table 2. Important competence to be a leader

School Principals	What Is/Are Important to Be a Leader?
P1	Working and competition
P2	Being objective, having no bias, being coherent, communicating with all colleagues in the same way
P3	Being a good listener and solution oriented
P4	Having team spirit and giving importance to colleagues and power of communicating
P5	Being honest and protective, increasing confidence
P6	The quality of communicating with people, being experienced, well-educated, and sophisticated
P7	Being determined, having the ability of thinking scientifically, having international ethics, being respectful to human right and fair
P8	The ability of communicating, having knowledge or experiences related to his/her own job
P9	Having vision, communicating, authority, motivating others.
P10	Being honest, determined, communication skills and making consensus
P11	Determined, strong-minded, daring
P12	Creating vision and mission, having communication skills
P13	Successful school administrator
P14	Determined, fair, self-confidence, empathy
P15	Determined, self-confidence, modesty and flexible
P16	Patient and sophisticated
P17	Thinking all the variation without causing offence, working with solution focused working, mutual respect and love
P18	Communication
P19	The feature of administrator
P20	The ability of problem solving and esteemed

consensus, solution oriented, well-educated, patient, the ability of thinking are expressed as important to be a leader.

The answer for the question of what makes it easy to be a leader is given in Table 3.

As seen in Table 3, communication is one of the frequently stated features that make being a leader easy. Establishing strong lines of communication with colleagues is frequently expressed by the principals as important and also as necessary, therefore communication is proved to be a pivotal skill that leaders have. Communication highlights the fact that leadership can be shared amongst multiple actors and relies on interrelationship between leaders and followers (Murphy, Elliott, Goldring, & Porter, 2006). A successful leader needs followers and if not making

Table 3. What makes it easy to be leader?

School Principals	What Makes It Easy to Be a Leader?
P1	Communication
P2	Treating equally to all, motivating, directing colleagues to the same goals
P3	Being calm and constructive
P4	Learning easy related to laws and regulations and implementing them
P5	Talented, problem solving and entrepreneur
P6	Communication
P7	Communicating effectively
P8	Success, knowing what to do and how to do, attracting people, attracting the team
P9	Communication, working hard and motivation
P10	The art of speaking in public eloquently and effectively, giving importance to others' ideas, taking risk
P11	Persuasion, determined, self-confidence, fair, leading, knowing colleagues well, recognizing the circumstances
P12	Honest, hard-working, communicating effectively
P13	Communication, master of his/her own job
P14	Communication skills, knowing the law
P15	Coordination and communication
P16	Finding solution for the problems, being a master of his/her own job
P17	Taking quick decision for the problems waiting, finding practical solutions, making colleagues feel how you value them, and making efforts to be well-educated. A master of his/her own business, being perfect in every field
P18	Thinking differently, seeing difference, the ability of thinking difference that will enable the changes, experiences
P19	Having experiences
P20	Attracting

an effective communications with each other, none of the partner benefits from the leadership practices. It is obvious that most of the principals have in mind the same view of their leadership that taking decision, being a master of his/her own business and the ability of problem solving, attracting others and objectivity make it easy to be a leader. Knowing the law and regulations, objectiveness, being different from the others are also seen as facilitating to be a leader.

Principals are asked to explain the reasons of not having problems in leading colleagues and the expression are given in Table 4.

Table 4 presents the reasons of not having problems in leading colleagues. These reasons expressed by principals indicate that not having problems in leading col-

Table 4. The reasons of not having problems in leading colleagues

School Principals	Why Do You Think You Do Not Have Difficulties in Leading Your Colleagues?
P1	They want innovations and development
P2	I know how well I am at the art of speaking in public eloquently and effectively and having a great empathy
P3	My colleagues know we are a team we do all the things together.
P4	I believe in my charisma and the my communication skills
P5	We all work together having the same goals
P6	I have a team spirit
P7	I have empathy with my colleagues and communication skills.
P8	I try not to see other's faults. I can criticize my colleagues without offending. I always keep in mind that we are human beings.
P9	I know how to persuade others.
P10	..
P11	I know my colleagues well and I am sincere
P12	I care my colleagues, give importance to their opinions and respect to their decisions
P13	I try to understand them.
P14	I always try to communicate with them and I also know what they want.
P15	We all share our responsibilities.
P16	I make my colleagues be involved in decision making process.
P17	I have empathy. I know my colleagues well and try to help them and I also try to make them feel this.
P18	I have good communication skills; I am receptive to their opinions. My colleagues can easily talk with me. We can communicate effectively.
P19	I think I express myself
P20	I can persuade my colleagues about why they are with me

leagues is related to the competence and skills that principals have. The majority of the principals have indicated an agreement with the statement that knowing each other, communication and empathy prevent having problems with colleagues. Principals also agreed that knowing how to communicate and showing empathy in any circumstances prevents troubles and solves the problems. Team spirit is also another emphasized reason of not having problem during leadership.

The main problems that principals have had are given in Table 5.

Table 5 indicates that principals mostly have various problems. They frequently express the problems related to personal characteristics or stemming from themselves such as losing control of anger, being yes-man, having no actual plans, ignorance

Table 5. Main problems during leadership

School Principals	What Are the Main Problems You Have during Your Leadership?
P1	Losing the control of my anger. I had many problems after this and spent time then to recover.
P2	Even though everything is going well, the achievement of the school hasn't reached its goals. This failure makes me feel how inefficient I am.
P3	The people who think themselves as a leader and having the ability to be a leader.
P4	I sometimes cannot find solution to personal problems.
P5	Being misunderstood
P6	My colleagues do not put their shoulder to the wheel.
P7	Ignorance and biases
P8	I do not illuminate my colleagues with my knowledge. I feel myself inefficient to deal with the conflicts.
P9	--
P10	When you ignore your colleagues.
P11	Being gentle and not saying no to my colleagues.
P12	When my colleagues have hesitation in my being sincere
P13	Working with people who do not understand each other and thinking themselves just.
P14	Being idealist
P15	Not making plans
P16	--
P17	Environmental problems which we do not interfere in effects our efficiency.
P18	I am misunderstood by my colleagues because of my types of communications and behaviors. This miss understanding creates trouble.
P19	I have difficulty in managing people.
P20	Personality conflict (clinical level), union and politics pressure

and biases. This list can be multiplied according to the circumstances. Still there are common ground reasons such as misunderstanding, ignorance and personal conflicts stated by principals.

CONCLUSION AND DISCUSSION

In this part, the main conclusions and recommendations of the research considering the analysis of data are presented. It is crucial to mention that the conclusions and recommendations cannot be freely generalizable since the research was done with a small study group in the city center of Ankara, Turkey. Nevertheless, it can be useful and relevant to other schools in Ankara, Turkey which have similar educational context. We do not seek to generalize findings from the cases, but rather to employ these as particular and to reveal windows into prevailing successful principal leadership in today's complex school. Through these windows, we hope to gain insight into the nature of school leadership emerging in complexity in the educational environment. This research on school leaders, with its quantitative data defining the characteristics of principal based on leadership in complexity may be just a mirror that principals need.

This study indicates that some main focused characteristics are captured from the principals' expression about their own leadership in complexity. These characteristics are listed below:

- Open communication,
- Shared decision,
- Consideration,
- Servant leader,
- Shared vision
- Not being afraid of making mistakes,
- Positive interpersonal skills,
- Growth-oriented.

This study also illustrates that educational leadership changes include a shift from the traditional system to the leader of complexity. In order to emphasize this change, principals stated some vital features listed below:

- Interconnectedness,
- Effective communication,
- Shared decision-making,
- Setting common goals,
- Having empathy,
- Knowing colleagues well,
- Give chances to teacher for professional development,
- Help teachers to solve problems together more effectively.

The study also shows that principals have some problems during leadership while managing the complexity. The reasons for these problems are:

- Misunderstanding,
- Not reaching the goals,
- Bias,

- Ignorance,
- Personal conflicts.

The findings from this study indicate that an effective leader of complexity at schools is one who is a facilitator rather than a controller. The leader as a facilitator is willing to share in;

- Decision-making process,
- Treats all educational shareholders with respect,
- Promote a collective vision,
- Frequently practices open communication and
- Encourage personal and professional growth.

This study gives overarching expressions for the principal leadership in complex schools. What these expressions had in common is that principals review the leadership as a complex and changing situation that includes many different behaviors. What the main conclusion of these interviews is that the principals consider the leaders in complex situations as a facilitator rather than a controller and emphasized that the leadership becomes meaningful when followers and leaders all work together underpinning team spirit. This inference through self-assessment of principals reminds us the importance of collective, distributed, instructional (Barton, 2013) and shared leadership (Tobin, 2014). This also indicates that again situations has moved from an early focus on formal leadership, on particular characteristics (Erçetin, Potas, Kısa & Açıkalın, 2013; Hard & Jónsdóttir, 2013; Yılmaz Fındık, 2015) through behaviors of leaders, complexity leadership (Erçetin, 2000; Erçetin & Kamacı, 2008; Erçetin & Kayman, 2011; Fris & Laridou, 2006; Hannah, Balthazard, Waldman, Jennings & Tatcher, 2013; Hard & Jónsdóttir, 2013; Morrison, 2002). Leaders often operate in complex situations in which they have to make decisions in this changing environment (Hannah, Balthazard, Waldman, Jennings & Tatcher, 2013, 393). These changing situations create uncertainty and unpredictability (Erçetin & Kamacı, 2008; Hannah, Balthazard, Waldman, Jennings & Tatcher, 2013) about what leaders are supposed to do and what leaders are required to negotiate to achieve effectiveness and school success. Not always one types of leadership is effective at school but many of them and in actual fact a mixture of the leadership characteristics, which can easily challenge with the changing and evolving situations and can meet diverse role demands at school, are effective and essential at school.

FUTURE TRENDS

School leaders have faced various changes to cope with at school and this makes school leadership evolve to deal with the complex changes to meet the demands of 21st century. The expectations of what school leaders should achieve has already change so the definition of school leaders must be change. This rapidly changing situation gives rise in complexity leadership and its role in social sciences and so education. It has become increasingly apparent that complexity leadership becomes well suited for the needs of 21st century and has been a new era to meet these changing dynamics.

Some limitations of this study deserve note; one is that sample and instrumentation focused on a small group and this prevents from generalizing the findings to a large group of principals. Number of study group and diversity of the questions can be multiplied. Future research can use different data collection methods such as observation and focus group discussion in addition to quantitative data collection. Future research should also recognize the need to apply research to a variety of educational environments to compare and contrast to a broader understanding of how principals assess their leadership process. The relationship between principal leadership and student success should also be investigated in the future research.

REFERENCES

Barton, L. T. (2013). Knowledge of effective educational leadership practices. *International Journal of Educational Leadership Preparation, 8*(1).

Blandford, S. (2013). The impact of 'achievement for all' on school leadership. *Educational Management Administration & Leadership*, *41*(1), 45–62. doi:10.1177/1741143212462701

Erçetin, Ş. Ş. (2000). *Lider Sarmalında Vizyon.* Ankara: Nobel Publications.

Erçetin, Ş. Ş., & Kamacı, M. C. (2008). Quantum Leadership Paradigm. *World Applied Sciences Journal*, *3*(6), 865–868.

Erçetin, Ş. Ş., Potas, N., Kısa, N. & Açıkalın, Ş. N. (2013). To be on the edge of chaos with organizational intelligence and health. In *Chaos and Complexity Theory for Management: Nonlinear Dynamics.* IGI Global. DOI: 10.4018/978-1-4666-2509-9.ch005

Fris, J., & Laridou, A. (2006). An additional way of thinking about organizational life and leadership: The quantum perspective. *Canadian Journal of Educational Administration and Policy*, 48.

Gomez, D. (2013). Leadership behavior and its impact on student's success and retention in online graduate education. *Academy of Educational Leadership Journal*, *17*(2), 13–37.

Hallinger, P. (2010). Developing Insructional Leadership. In Developing Successful Leadership, Studies in Educational Leadership 11. Springer Science+Business Media B.V.

Hannah, S. T., Balthazard, P. A., Waldman, D. A., Jennings, P. L., & Tatcher, R. W. (2013). The psychological and neurological bases of leader self-complexity and effects on adaptive decision-making. *The Journal of Applied Psychology*, *98*(3), 393–411. doi:10.1037/a0032257 PMID:23544481

Hard, L., & Jónsdóttir, A. H. (2013). Leadership is not a dirty word: Exploring and embracing leadership in ECEC. *European Early*, *21*(3), 311–325. doi: 10.1080/1350293X.2013.814355

Henning, J. E., & Robinson, V. (2002). Change and transformation inaneducational leadership program. In Changing World of School Administration. National Council of Professors of Educational Administration.

Kayman, E. A., & Erçetin, Ş. Ş. (2011). The level of trained school principals' fulfillment of quantum leadership behaviors within Strengthening Vocational Education and Training (SVET) Project in Turkey. *Humanity and Social Sciences Journal*, *6*(1), 16–21.

Klar, H. W. & Brewer, C. A. (2013). Successful leadership in high-needs schools: an examination of core leadership practices enacted in challenging contexts. *Educational Administration Quarterly*, *49*(5), 768–808. DOI: 10.1177/0013161X13482577

Leithwood, K., Louis, K. S., Anderson, S., & Wahlstrom, K. (2004). *How leadership influences student learning (paper commissioned by the Wallace Foundation)*. Minneapolis, MN: University of Minnesota.

Leithwood, K., Mascall, B., & Strauss, T. (2009). *Distributed Leadership According to the Evidence*. London: Rouhledge.

Marzano, R., Waters, T., & McNulty, B. (2005). *School leadership that works: from research to results*. Aurora, CO: Mid-Continent Research for Education and Learning.

McCarthy, M. M. (2002). The changing environment for school leaders: market forces. In Changing World of School Administration. National Council of Professors of Educational Administration.

McEwan, E. (2003). *10 Traits of Highly Effective Principals: From Good To Great Performance*. Thousand Oaks, CA: Corwin Press, Inc.

Morrison, K. (2002). *School Leadership and Complexity Theory*. New York: Routledge.

Morrison, K. (2015). Complexity theory, school leadership and management: Questions for theory and practice. *Educational Management Administration & Leadership*, *38*(3), 374–393. doi:10.1177/1741143209359711

Murphy, J., Elliot, S., Goldring, E., & Porter, A. (2006). *Learning-Centered Leadership: A Conceptual Foundation*. Paper presented at the annual meeting of the Wallace Foundation State Action for Educational Leadership Conference, Saint Louis, MO.

Murphy, J., Elliot, S., Goldring, E., & Porter, A. (2007). Leadership for learning: a research-based model and taxonomy of behaviors. *Journal of School Leadership and Management, 27*(2), 179–201. Retrieved November, 3 2014, from http://www.valed.com/documents/3_murphy%20et%20al_2007.pdf

Murphy, J., & Hallinger, P. (1988, February). The characteristics of instructionally effective school districts. *The Journal of Educational Research*, *81*(3), 176–181. doi:10.1080/00220671.1988.10885819

Ogawa, R. T., & Bossert, S. T. (1995). Leadership as an organizational quality. *Educational Administration Quarterly*, *31*(2), 224–243. doi:10.1177/0013161X95031002004

Raja, A. S. & Palanichamy, P. (2011). Leadership styles of school principals- an examination at Madurai district. *Global Management Review, 5*(4).

Robinson, V., Lloyd, C., & Rowe, K. (2008). The impact of leadership on student outcomes: An analysis of the differential effects of leadership types. *Educational Administration Quarterly*, *44*(5), 635–674. doi:10.1177/0013161X08321509

Tillman, L. C. (2002). The impact of diversity in educational administration. In Changing World of School Administration. National Council of Professors of Educational Administration.

Tobin, J. (2014). Management and leadership issues for school building leaders. *International Journal of Educational Leadership Preparation, 9*(1).

Uhl-Bien, M., Marion, R., & McKelvey, B. (2007). *Complexity Leadership Theory: Shifting Leadership from the Industrial Age to the Knowledge Era*. Lincoln, NE: Leadership Institute Faculty Publications.

Wahlstrom, K. L., Louis, K. S., Leithwood, K., & Anderson, S. E. (2010). *Learning from Leadership: Investigating the Links to Improved Student Learning*. St. Paul, MN: University of Minnesota.

Ward, C. J. (2013). Why leadership matters: one school's journey to success. *Educational Leadership and Administration: Teaching and Program Development, 24*, 62-73.

Weng, C.-H., & Tang, Y. (2014). The relationship between technology leadership strategies and effectiveness of school administration: An empirical study. *Computers & Education, 76*, 91–107. doi:10.1016/j.compedu.2014.03.010

Yılmaz Fındık, L. (2015). Postgraduate students' perceptions of leadership in politics: a metaphor analysis sample of Hacettepe university, department of educational sciences. In Ş.Ş. Erçetin & S. Banerjee (Eds.), Chaos, Complexity and Leadership 2013. Springer. DOI doi:10.1007/978-3-319-09710-7_47

ADDITIONAL READING

Banerjee, S., & Erçetin, Ş. Ş. (2014). *Chaos, Complexity and Leadership 2012, Springer Proceedings in Complexity*. Springer; doi:10.1007/978-94-007-7362-2

Erçetin, Ş. Ş., & Banerjee, S. (2014). *Chaos and Complexity Theory in World Politics*. USA: IGI Global. doi:10.4018/978-1-4666-6070-0

Erçetin, Ş. Ş., & Banerjee, S. (2015). *Chaos, Complexity and Leadership 2013. Springer Proceedings in Complexity*. New York: USA Springer Understanding Complex Systems Series.

Erçetin, Ş. Ş., Potas, N., Kısa, N. and Açıkalın, Ş. N. (2013). To Be On The Edge of Chaos with Organizational Intelligence and Health. Chaos and Complexity Theory for Management: Nonlinear Dynamics. IGI Global. 89-104. DOI: 10.4018/978-1-4666-2509-9.ch005

Morrison, K. (2015). Complexity Theory, School Leadership and Management: Questions For Theory and Practice. *Educational Management Administration & Leadership, 38*(3), 374–393. doi:10.1177/1741143209359711

Uhl-Bien, M., Marion, R., & McKelvey, B. (2007). *Complexity Leadership Theory: Shifting Leadership from the Industrial Age to the Knowledge Era*. Lincoln: Leadership Institute Faculty Publications.

KEY TERMS AND DEFINITIONS

Complexity Leadership: A new paradigm of the leadership process that enables quality outcomes and innovation produces appropriate vision and mission for the organization and based on complex relationship of many interactive factors and team performance.

Complexity: Various unpredicted changes are emergent.

Leadership: The process of influencing the behavior of follower in order to accomplish the goals of an organization.

School Leader: Leader who creates positive educational environment for both teachers and students for learning in order to ensure quality education.

Self-Assessment: Assessing his/her own process and making judgements.

Compilation of References

Abebe, M., Angriawan, A., & Tran, H. (2010). Chief executive external network ties and environmental scanning activities: An empirical examination. *Strategic Management Review*, *4*(1), 30–43.

Abraham, C., & Sheeran, P. (2003). Acting on intentions: The role of anticipated regret. *The British Journal of Social Psychology*, *42*(4), 495–511. doi:10.1348/014466603322595248 PMID:14715114

Abraham, R. R., Vinod, P., Kamath, M. G., Asha, K., & Ramnarayan, K. (2008). Learning approaches of undergraduate medical students to physiology in a non PBL and partially PBL- oriented course. *Advances in Physiology Education*, *32*(1), 35–37. doi:10.1152/advan.00063.2007 PMID:18334566

Abu-Doleh, J., & Weir, D. (2007). Dimensions of performance appraisal systems in Jordanian private and public organizations. *International Journal of Human Resource Management*, *18*(1), 75–84. doi:10.1080/09585190601068334

Acevedo, M., & Krueger, J. I. (2004). Two egocentric source of the decision to vote: The voter's illusion and the belief in personal relevance. *Political Psychology*, *25*(1), 115–134. doi:10.1111/j.1467-9221.2004.00359.x

Açıkalın, Ş. N., & Bölücek, C. A. (2014). Understanding of Arab Spring with chaos theory-uprising or revolution. In S. Banerjee (Ed.), Chaos theory in politics (pp. 29-47). Springer.

Adepoju, T. L. (2003a). *Trade unionism and job productivity in human organization*. Ibadan: Hope Publications.

Adepoju, T. L. (2003b). *Trade unionism and culture of violence and resistance: A social system perspective*. Ibadan: Hope Publications.

Afshari, M., Bakar, K. A., Luan, W. S., Samah, B. A., & Fooi, F. S. (2009). Technology and school leadership. *Technology, Pedagogy and Education*, *18*(2), 235–248. doi:10.1080/14759390902992527

Afzahur, R. M. (2001). *Managing conflict in organizations* (3rd ed.). Greenwood.

Ajan, H., & Hartshorne, R. (2008). Investigating faculty decisions to adopt Web 2.0 technologies: Theory and empirical tests. *The Internet and Higher Education*, *11*(2), 71–80. doi:10.1016/j.iheduc.2008.05.002

Albion, P. (2008). Web 2.0 in teacher education: Two imperatives for action. *Computers in the Schools*, *25*(3/4).

Alfred, M. (2003). Kleptocracy and divide and rule: A model of personal rule. Paper presented at the Marshal European economic association annual lecture, Stockholm, Sweden.

Alibi, A. T. (n.d.). *Conflict in Nigeria University: cause and management*. Academic Press.

Alıcıgüzel, İ. (1979). *Education in first and middle degree schools*. Istanbul: İnkılap ve Aka Bookstore.

Alifa, D. (2001, June 17). Intrigues in FG-ASUU Face-off. *The Guardian*.

Allen, P. M. (2001). A complex systems approach to learning in adaptive networks. *International Journal of Innovation Management*, *5*(02), 149–180. doi:10.1142/S136391960100035X

Altunoğlu, S. (2009). Kaosun Keşfi ve Felsefesi, Bibliotech. *Felsefe-Sosyal Bilimler Dergisi, Eylül/Ekim* (9).

Amin, E. M. (2005). *Social science research: Conception, methodology and analysis*. Kampala: Makerere University Press.

Amrein-Beardsley, A., Berliner, D. C., & Rideau, S. (2010). Cheating in the first, second, and third degree: Educators' responses to high-stakes testing. *Education Policy Analysis Archives*, *18*(14), 36.

Anderson, J. W., Foster-Kehn, M., & McKinney, B. C. (1996). *Communication skills for surviving conflicts at work*. New Jersey: Hampton Press.

Anderson, S., & Wahlstrom, K. (2004). *How leadership influences student learning*. New York: The Wallace Foundation.

Arroba, T. (1998). Decision making by Chinese – *US. The Journal of Social Psychology*, *38*, 102–116.

Ashford, S. J. (1986). Feedback-seeking in individual adaptation: A resource perspective. *Academy of Management Journal*, *29*(3), 465–487. doi:10.2307/256219

Ashford, S. J., & Cummings, L. L. (1983). Feedback as an individual resource: Personal strategies of creating information. *Organizational Behavior and Human Performance*, *32*(3), 370–398. doi:10.1016/0030-5073(83)90156-3

Ashford, S. J., Rothbard, N. P., Piderit, S. K., & Dutton, J. E. (1998). Out on a limb: The role of context and impression management in selling gender-equity issues. *Administrative Science Quarterly*, *43*(1), 23–57. doi:10.2307/2393590

Ashforth, B. E., & Humphrey, R. H. (1995). Emotion in the workplace: A reappraisal. *Human Relations*, *48*(2), 97–125. doi:10.1177/001872679504800201

Ashton, P. T., & Rodman, B. W. (1986). *Making a difference: Teachers' sense of self-efficacy and student achievement* (p. 45). New York: Longman publishers.

Atherton, J. S. (2005). Learning and teaching: Learning how to learn. Academic Press.

Atherton, J. S. (2008). Doceo: Teacher-centred teaching works. Academic Press.

Atmosudirdjo, P. (1987). *Beberapa pandangan umum tentang pengambilan keputusan*. Jakarta: Ghalia Indonesia.

Balcı, A. (2001). *Etkili okul ve okul geliştirme*. Ankara: PegemA Yayıncılık.

Balcı, A. (2010). *Sosyal bilimlerde araştırma yöntem teknik ve ilkeler (7. baskı)*. Ankara: Pegem Akademi.

Balcı, A. (2010). *Sosyal bilimlerde araştırma: Yöntemler, teknikler, ilkeler*. Ankara: PegemA Yayıncılık.

Balcı, A. (2015). *Sosyal bilimlerde arastirma: Yöntem, teknik ve ilkeler. (10'th Edition)*. Ankara: PegemA Yayincilik.

Balcı, A., Baltacı, A., Fidan, T., Cereci, C., & Acar, U. (2012). Örgütsel sosyalleşmenin, örgütsel özdeşleşme ve örgütsel vatandaşlıkla ilişkisi: Ilköğretim okulu yöneticileri üzerinde bir araştırma. *Eğitim Bilimleri Araştırmaları Dergisi-Journal of Educational Sciences Research*, *2*(2), 47–74.

Banks, J., & Smyth, E. (2015). Your whole life depends on it: Academic stress and high-stakes testing in Ireland. *Journal of Youth Studies*, *18*(5), 598–616. doi:10.1080/13676261.2014.992317

Barner-Rasmussen, W. (2003). Determinants of the feedback-seeking behaviour of subsidiary top managers in multinational corporations. *International Business Review*, *12*(1), 41–60. doi:10.1016/S0969-5931(02)00087-2

Bar-On, R. (2006). The Bar-On model of emotional-social intelligence (ESI). *Psicothema*, *18*(1), 13–25. PMID:17295953

Barton, L. T. (2013). Knowledge of effective educational leadership practices. *International Journal of Educational Leadership Preparation*, *8*(1).

Bass, B. M., & Bass, R. (2009). *The Bass handbook of leadership: Theory, research, and managerial applications*. Simon and Schuster.

Bateman, T. S., & Crant, J. M. (1993). The Proactive Component of Organizational Behavior: A Measure and Correlates. *Journal of Organizational Behavior*, *14*(2), 103–118. doi:10.1002/job.4030140202

Bateman, T. S., & Crant, J. M. (1999). Proactive behavior: Meaning, impact, recommendations. *Business Horizons*, *42*(3), 63–70. doi:10.1016/S0007-6813(99)80023-8

Bate, S. P., & Robert, G. (2002). Knowledge management and communities of practice in the private sector: Lessons for modernizing the National Health Service in England and Wales. *Public Administration*, *80*(4), 643–663. doi:10.1111/1467-9299.00322

Bear, G. (2011). *Discipline: Effective school practices. National Association of School Psychologist*. Newark, NJ: University of Delaware.

Becta. (2009). *Literature Review: Deep learning with technology in 14- to 19-year-old learners*. Retrieved from http://publications.becta.org.uk/display.cfm?resID=31532&page=1835

Berardi, A., & Blackmore, C. (2006). *Introducing environmental decision making*. Open University Worldwide.

Bernard, C. (1938). *The functions of the executive*. Cambridge, MA: Harvard University press.

Betts, J. R., & Grogger, J. (2000). *The impact of grading standards on student achievement. Educational attainment and entry-level earnings*. NBER Working Paper 7875.

Betts, J. R. (1995). *Do grading standards affect the incentive to learn? Manuscript, University of California*. San Diego, CA: Department of Economics.

Beyer, J., & Niño, D. (2001). Culture as a source, expression, and reinforcer of emotions in organizations. In R. L. Payne & C. L. Cooper (Eds.), *Emotions at work: Theory, research, and applications for management* (pp. 173–197). Chichester, UK: Wiley.

Biggs, J. (1987). *Student approaches to learning and studying*. Hawthorn, Australia: Australian Council for Educational Research.

Bilge, A., & Çam, O. (2010). Ruhsal hastalıklara yönelik damgalama ile mücadele. *TAF Preventive Medicine Bulletin*, *9*(1), 71–78.

Bindl, U., & Parker, S. K. (2010). *Proactive work behavior: Forward-thinking and change-oriented action in organizations* (Vol. 2). Washington, DC: American Psychological Association.

Bisaso, S. M., & Rafiyyah, S. A. (2010). *The relationship between perception of learning and teaching practices of lecturers: A study of Educators at IIUM*. Malaysian Research Repository.

Bisaso, S. M., & Rafiyyah, S. A. (2010). *The Relationship between perception of learning and teaching practices of lecturers: A study of educators at IIUM*. Malaysian Research Repository.

Bishop, K., Webber, S. S., & O'Neill, R. (2011). Preparation and prior experience in issue-selling success. *Journal of Managerial Issues*, 323–340.

Black, R. (2007). *Overcoming disadvantage through the innovative classroom*. Education Foundation Australia. Retrieved from http:www.educationfoundation.org.au

Blandford, S. (2013). The impact of 'achievement for all' on school leadership. *Educational Management Administration & Leadership*, *41*(1), 45–62. doi:10.1177/1741143212462701

Blase, J. J. (1990). Some Negative Effects of Principals' Control-Oriented and Protective Political Behaviour. *American Educational Research Journal*, *27*(4), 727–753. doi:10.3102/00028312027004727

Blass, F. R., Brouer, R. L., Perrewé, P. L., & Ferris, G. R. (2007). Politics Understanding and Networking Ability as a Function of Mentoring The Roles of Gender and Race. *Journal of Leadership & Organizational Studies*, *14*(2), 93–105. doi:10.1177/1071791907308054

Blazer, C. (2011). Unintended consequences of high-stakes testing. Miami-Dade County Public Schools.

Bloch, D. P. (2005). Complexity, chaos and nonlinear dynamics: A new perspective on career development theory. *The Career Development Quarterly*, *53*(3), 194–207. doi:10.1002/j.2161-0045.2005.tb00990.x

Blumer, H. (1986). *Symbolic Interactionism: Perspective and Method.* Berkeley, CA: University of California Press.

Borgatti, S. P., & Cross, R. (2003). A relational view of information seeking and learning in social networks. *Management Science*, *49*(4), 432–445. doi:10.1287/mnsc.49.4.432.14428

Brandt, R. (1987). On cooperation in schools: A conversation with David and Roger Johnson. *Educational Leadership*, *45*(3), 14–19.

Brass, D. J. (1985). Men's and women's networks: A study of interaction patterns and influence in an organization. *Academy of Management Journal*, *28*(2), 327–343. doi:10.2307/256204

Bregman, J., & Bryner, K. (2003). *Quality of secondary education in Africa (SEIA). Working African Development Bank, (AfDB)*. Human Development Indicators.

Brent, D. R., & Susan, J. (2012). *Leading in tough times: Case studies for higher education leaders*. NACUBO.

Brien, K. & Stelmach, B. L. (2009). Legal and cultural contexts of parent-teacher interactions: school councils in Canada. *International Journal about Parents in Education, 3*(1), 1-14.

Brosky, D. (2011). Micropolitics in the School: Teacher Leaders' Use of Political Skill and Influence Tactics. *International Journal of Educational Leadership Preparation*, *6*(1), n1.

Brown, J. A. (2004). *African American church goers' attitudes toward treatment seeking from mental health and religious sources: The role of spirituality, cultural mistrust and stigma toward mental illness.* (Unpublished doctoral dissertation). University of North Caroline, Chapel Hill, NC.

Brown, G. T. L. (2002). Student beliefs about learning: New Zealand students. *Academic Exchange Quarterly*, *6*(1), 110–114.

Brown, G. T. L., Lake, R., & Matters, G. (2008). New Zealand and Queensland teachers' conceptions of learning: Transforming more than reproducing. *Australian Journal of Educational & Developmental Psychology.*, *8*, 1–14.

Bruine de Bruin, W., Parker, A. M., & Fischhoff, B. (2007). Individual differences in adult decision making competence. *Journal of Personality and Social Psychology*, *92*(5), 938–956. doi:10.1037/0022-3514.92.5.938 PMID:17484614

Buchanan, J. M., & Devletiglou, N. E. (1971). *Academia in anarchy: An economic diagnosis.* London: Tom Stacey Ltd.

Burns, T. E., & Stalker, G. M. (1961). *The management of innovation.* London: Tavistock.

Bursalıoğlu, Z. (2013). *Okul yönetiminde yeni yapı ve davranış.* Ankara: Pegem Akademi.

Bush, T. (2003). *Theories of educational management* (3rd ed.). London: Sage Publishers.

Buyruk, H. (2008). *Eğitimde yaşanan toplumsal eşitsizliklere ilişkin biyografik bir araştırma. Yayımlanmamış Yüksek Lisans Tezi.* Ankara: Ankara Üniversitesi Eğitim Bilimleri Enstitüsü Eğitim Bilimleri Anabilim Dalı Eğitim Ekonomisi ve Planlaması Programı.

Cahill, K. (2010). *Even in chaos: Education in times of emergency.* Fordham University press.

Camara, N. J. (2011). *Life after disability diagnosis: The impact of special education labeling in higher education.* (Unpublished doctoral dissertation). Georgia State University, Atlanta, GA.

Campbell, D. J. (2000). The proactive employee: Managing workplace initiative. *The Academy of Management Executive*, *14*(3), 52–66.

Candelaria, M. L., & Limjap, A. A. (2002). Problem solving heuristics of college freshmen: A case analysis. *Asia Pacific Education Review*, *11*(2), 199–223.

Canter, L., & Canter, M. (1976). *Assertive discipline: A take-charge approach for today's educator.* Los Angeles, CA: Canter and Associates.

Capogna, S. (2010). Learning and social networks: New forms of sociability. In *ESA Research Network Sociology of Culture Midterm Conference: Culture and the Making of Worlds.*

Carli, L. L. (2001). Gender and social influence. *The Journal of Social Issues*, *57*(4), 725–741. doi:10.1111/0022-4537.00238

Cartwright, D., & Zander, A. (1958). *Group Dynamics.* Evanston, IL: Row-Peterson.

Charles, C. M. (2002). *Building classroom discipline* (6th ed.). New York: Longman.

Chiang, T. L., & Wang, J. (2008). *Workplace Learning, Job Satisfaction, and Organizational Commitment in Small to Midsize Companies in Taiwan.* Online Submission, retrieved on 15.05.2015.

Choo, C. W. (2001). Environmental scanning as information seeking and organizational learning. *Information Research*, *7*(1), 7–1.

Choudhury, V., & Sampler, J. L. (1997). Information specificity and environmental scanning: An economic perspective. *Management Information Systems Quarterly*, *21*(1), 25–53. doi:10.2307/249741

Churchill, D. (2009). Educational applications of Web 2.0: Using blogs to support teaching and learning. *British Journal of Educational Technology*, *40*(1), 179–183. doi:10.1111/j.1467-8535.2008.00865.x

Clapham, C., & Nicholson, J. (2009). *The concise Oxford dictionary of mathematics*. Oxford University Press. doi:10.1093/acref/9780199235940.001.0001

Clark, B. (1998). Creating entrepreneurial universities: organizational pathways of transformation. Perganon, UK: IAU Press.

Çobanoğlu, F. (2008). Değişim mantığını anlamak: Akış ve dönüşüm olarak örgüt. *Pamukkale Üniversitesi Eğitim Fakültesi Dergisi*, *1*(23), 110–119.

Cobbold, T. (2009). *Local area school comparisons are 'simplistic' and unfair league tables*. Academic Press.

Cohen, L., Manion, L., & Morrison, K. (2007). *Research Methods in Education* (6th ed.). New York: Routledge.

Cole, G. A. (1999). *Management: Theory and practice (5th ed.)*. London: Letts Educational Aldine Place.

Coleman, B., & McNeese, M. N. (2009). From home to school: The relationship among parental involvement, student motivation, and academic achievement. *The International Journal of Learning*, *16*(7), 459–470.

Coleman, J., Hoffer, T., & Kilgore, S. (1982). Cognitive outcomes in public and private schools. *Sociology of Education*, *55*(2/3), 65–76. doi:10.2307/2112288

Coloroso, B. (2002). *Kids are worth it! Giving your child the gift of inner discipline*. New York: Harper Collins.

Cooper, R. K. & Sawaf, A. (1998). Executive EQ: emotional intelligence in leadership and organizations. New York: Berkley Publishing Group.

Cooper, S. M. (2002). Classroom choices for enabling peer learning. *Theory into Practice*, *41*(1), 53–57. doi:10.1207/s15430421tip4101_9

Corey, G., & Corey, M. (1982). *Groups: Process and practice*. Monterey, CA: Brooks/Cole.

Corvette, B. A. B. (2007). *Conflict management: A practical guide to developing negotiation strategies*. Prentice Hall.

Crant, J. M. (1996). The Proactive Personality Scale as a Predictor of Entrepreneurial Intentions. *Journal of Small Business Management*, *34*(3), 42–49.

Crant, J. M. (2000). Proactive Behavior in Organizations. *Journal of Management*, *26*(3), 435–462. doi:10.1177/014920630002600304

Credaro, A. (2006). Innovation and Change in Education. *Worrier Librarian*. Retrieved from www.warriorlibrarian.com/LIBRARY/innovate.html

Cremona, K. (2012). *A framework for understanding organisational dynamics*. Australian Businesswomen's Network. Available from: http://www.abn.org.au/site/article/organizational-dynamics

Cunningham, R. (2000). *Chaos, complexity and the study of education communities*. Paper presented to the British Educational Research Association Annual Conference, University of Leeds.

Cunningham, R. (2000). *Chaos, complexity and the study of education communities*. Institute of Education.

Daft, R., & Marcic, L. D. (2001). Understanding management. San Diego, CA: Harcourt College Publisher.

Daft, R. (2014). *The leadership experience*. Cengage Learning.

Dalkıranoğlu, T., & Çetinel, F. G. (2008). Konaklama işletmelerinde kadın ve erkek yöneticilerin cinsiyet ayrımcılığına karşı tutumlarının karşılaştırılması. *Dumlupınar Üniversitesi Sosyal Bilimler Dergisi, 20*, 277–297.

Daniel, A., & Moudic, Y. (2010). *Female and gender leadership: To which extents gender impacts on leadership and organizations in France?* (Master dissertation). Linnaeus University.

De Bot, K., Verspoor, M., & Lowie, W. (2005). Dynamic systems theory and applied linguistics. *International Journal of Applied Linguistics, 15*(1), 116–118. doi:10.1111/j.1473-4192.2005.0083b.x

Definition of formal organization. (n.d.). Retrieved on 21st February, 2014 from www.businessdictionary.com/definition/formal-organisation.html

DeLamater, J. D., Myers, D. J., & Collett, J. L. (2015). Social Psychology (8th ed.). Boulder, CO: Westview Press.

Demchenko, Y. (1998). Paradigm Change in Education in Conditions of Emerging New Information Technologies and Global Information Infrastructure Building. *ED-Media/ED-Telecom 98 Proceedings*.

Demir, M. K., & Ari, E. (2013). Öğretmen sorunlar- Çanakkale ili örneği. *Ondokuz Mayıs University Education Faculty Magazine*, *32*(1), 10.

Dike, V. (2001). *Who is the author of democracy and political life in Nigeria*. Zaria: Ahmadu Bello University Press.

DiTomaso, N., & Farris, G. F. (1992). Diversity in the high-tech workplace: diversity and performance in R&D. *Spectrum, IEEE, 29*(6), 21-24.

Dönmez, B., & Sincar, M. (2008). Avrupa birliği sürecinde yükselen ağ toplumu ve eğitim yöneticileri. *Electronic Journal of Social Sciences*, *7*(24), 1–19.

Douglas, C., & Ammeter, A. P. (2004). An examination of leader political skill and its effect on ratings of leader effectiveness. *The Leadership Quarterly*, *15*(4), 537–550. doi:10.1016/j.leaqua.2004.05.006

Downing, K., Chan, S. W., Downing, W. K., Kwong, T., & Lam, T. F. (2008). Measuring gender differences in cognitive functioning. *Multicultural Education & Technology Journal*, *2*(1), 4–18. doi:10.1108/17504970810867124

Duncan, D. (2002). Emotionally intelligent female primary head teachers: The new women of power? *Education*, *30*(3), 48–54.

Duncker, K. (1945). On problem solving. *Psychological Monographs, 58*(5).

Dutton, J. E., & Ashford, S. J. (1993). Selling issues to top management. *Academy of Management Review*, *18*(3), 397–428.

Dutton, J. E., Ashford, S. J., O'Neill, R. M., & Lawrence, K. A. (2001). Moves that matter: Issue selling and organizational change. *Academy of Management Journal*, *44*(4), 716–736. doi:10.2307/3069412

Edutopia. (2008). *Why Intergrate Technilogy into the curriculum? The Reasons Are Many*. Available on www.edutopia.org

Eglene, O., Dawes, S. S., & Schneider, C. A. (2007). Authority and leadership patterns in public sector knowledge networks. *American Review of Public Administration, 37*(1), 91–113. doi:10.1177/0275074006290799

Ekici, M., & Kıyıcı, M. (2012). Sosyal Ağların Eğitim Bağlamında Kullanımı. *Uşak Üniversitesi Sosyal Bilimler Dergisi*, *5*(2), 156–167.

Elliott, G. C., Ziegler, H. L., Altman, B. M., & Scott, D. R. (1982). Understanding stigma: Dimensions of deviance and coping. *Deviant Behavior*, *3*(3), 275–300. doi:10.1080/01639625.1982.9967590

Ellis, C. W. (2005). *Management skills for new managers*. New York: AMACOM.

Entwistle, N., & McCune, V. (2004). The conceptual bases of study strategy inventories. *Educational Psychology Review*, *16*(4), 325–345. doi:10.1007/s10648-004-0003-0

Erçetin, Ş. Ş., & Maya, İ. (2005). Meslek-aile profilleri ile kadın ilköğretim okul müdürlerinin yönetsel görevler üstlenmelerinde karşılaştıkları engeller. 28-30 Eylül 2005 Denizli-Türkiye. XIV. Ulusal Eğitim Bilimleri Kongresi. Pamukkale Üniversitesi. Türkiye: Denizli.

Erçetin, Ş. Ş., Açikalin, S.N., & Bülbül, M.S. (2012). A multi-dimensional approach to leadership in chaotic environments. In S. Banerjee (Ed.), Chaos and Complexity Theory for Management: Nonlinear Dynamics: Nonlinear Dynamics (pp. 89-104). IGI Global.

Erçetin, Ş. Ş., Bisaso, S.M., & Saeed, F. (2015a). Understanding chaos and complexity in education systems through conceptualization of fractal properties. In Ş. Ş. Erçetin &S. Banerjee (Eds.), Chaos, Complexity and Leadership 2013 (pp. 147-161). Springer International Publishing.

Erçetin, Ş. Ş., Potas, N., Kısa, N., & Açıkalın, Ş.N. (2013). To be on the edge of chaos with organizational intelligence and health. In S Banerjee (Ed.), Chaos and complexity theory for management: Nonlinear dynamics (pp. 184–203). Hershey, PA: IGI Global.

Erçetin, Ş. Ş. (2000). *Lider Sarmalında Vizyon*. Ankara: Nobel Publications.

Erçetin, Ş. Ş. (2000). *Lider sarmalında vizyon*. Ankara: Nobel Yayınları.

Erçetin, Ş. Ş. (2015c). Understanding Recep Tayyip Erdoğan and Turkey with Plasma as a metaphor of the fourth state of matter. In Ş. Ş. Erçetin (Ed.), *Complexity and Leadership 2014* (pp. 1–13). Springer International Publishing.

Erçetin, Ş. Ş., Açıkalın, Ş. N., & Bülbül, M. Ş. (2013a). A Multi-Dimensional approach to leadership in chaotic environments. In S. Banerjee (Ed.), *Chaos and Complexity Theory for Management* (pp. 89–104). IGI Global. doi:10.4018/978-1-4666-2509-9.ch005

Erçetin, Ş. Ş., & Bisaso, S. M. (2015). The incorporation of fractals into educational management and its implications for school management models. In Ş. Ş. Erçetin & S. Banerjee (Eds.), *Chaos, Complexity and Leadership 2013*. Springer Proceedings in Complexity. doi:10.1007/978-3-319-09710-7_4

Erçetin, Ş. Ş., Bisaso, S. M., & Saeed, F. (2015). Understanding chaos and complexity in education systems through conceptualization of fractal properties. In Ş. Ş. Erçetin & S. Banerjee (Eds.), *Chaos, Complexity and Leadership 2013*. Springer Proceedings in Complexity. doi:10.1007/978-3-319-09710-7_12

Erçetin, S. S., & Düzer, A. (2008). Multidimensional perceptual leadership model. *World Applied Sciences Journal*, *3*(1), 25–33.

Erçetin, Ş. Ş., & Kamacı, M. C. (2008). Quantum Leadership Paradigm. *World Applied Sciences Journal*, *3*(6), 865–868.

Erçetin, Ş. Ş., & Kamacı, M. C. (2008). Quantum leadership paradigm. *World Applied Sciences Journal*, *3*(6), 865–868.

Erçetin, Ş. Ş., & Neyişçi, N. B. (2013a). Social networks: Connections in structures. In Ş. Ş. Erçetin & S. Banerjee (Eds.), *Chaos, Complexity and Leadership 2013*. Springer Proceedings in Complexity. doi:10.1007/978-3-319-09710-7_26

Erçetin, Ş. Ş., Potas, N., Kısa, N., & Açıkalın, Ş. N. (2013b). To be on the edge of chaos with organizational intelligence and health. In S. Banerjee (Ed.), *Chaos and Complexity Theory for Management* (pp. 184–203). IGI Global. doi:10.4018/978-1-4666-2509-9.ch009

Erden, M., & Akman, Y. (1998). *Education psychology*. Ankara: Arkadaş Press.

Ereş, F. (2006). Türkiye'de kadının statüsü ve yansımaları. *Gazi Üniversitesi Endüstriyel Sanatlar Eğitim Fakültesi Dergisi*, *19*, 40–52.

Ergün, M. (2011). Eğitim ve kalkınma. Dicle Üniversitesi Sosyal Araştırmalar Merkezi 3. Sosyal Bilimler Sempozyumu: Bölgesel Kalkınmada Eğitimin Rolü, 1-15. Diyarbakır: Sosyal Araştırmalar Merkezi.

Essien, J. E. (1975). *Education and challenges of quality*. Benin Ethiope Publishing Corporation.

European University Association. (2007). *Managing the university community: exploring good practice*. Retrieved from www. Eua.be

European University Association. (2007). *Managing the University Community: Exploring Good practice*. Retrieved on 25th mar, 2015 through www.eua.be

Eyal, O., & Kark, R. (2004). How do transformational leaders transform organizations? A study of the relationship between leadership and entrepreneurship. *Leadership and Policy in Schools*, *3*(3), 211–235. doi:10.1080/15700760490503715

Farenga, P. (2013). *Insularity of school research*. Retrieved from http://www.johnholtgws.com/pat-farengas-blog/2013/12/17/the-insularity-of-school-research

Ferreira, M. A. M., Filipe, J.A.C.B., Coelho, M., & Pedro, I.C. (2014). Chaos in World Politics: A Reflection. In S. Banerjee (Ed.), Chaos Theory in Politics (pp. 89-104). Springer.

Ferris, G. R., Treadway, D. C., Kolodinsky, R. W., Hochwarter, W. A., Kacmar, C. J., Douglas, C., & Frink, D. D. (2005). Development and validation of the political skill inventory. *Journal of Management*, *31*(1), 126–152. doi:10.1177/0149206304271386

Fhim, M. & Talabari, A.F. (2014). Chaos and complexity theory and education. *The Journal of English Language and Learning*, 13.

Figlio, D. N., & Lucas, M. E. (2003).*Do High Grading Standards Affect Student Performance?* University of Florida and National Bureau of Economic Research.

Foley, R. M. (2001). Professional development needs of secondary school principals of collaborative-based service delivery models. *High School Journal*, *85*(1), 10–23. doi:10.1353/hsj.2001.0016

Follet, M. P. (1940). *Papers on dynamic administration*. New York: Harper.

Frese, M., & Fay, D. (2001). 4. Personal initiative: An active performance concept for work in the 21st century. *Research in Organizational Behavior*, *23*, 133–187. doi:10.1016/S0191-3085(01)23005-6

Frese, M., Fay, D., Hilburger, T., Leng, K., & Tag, A. (1997). The concept of personal initiative: Operationalization, reliability and validity in two German samples. *Journal of Occupational and Organizational Psychology*, *70*(2), 139–161. doi:10.1111/j.2044-8325.1997.tb00639.x

Friedman, M. (1997). Public schools: Make them private. *Education Economics*, *5*(3), 341–344. doi:10.1080/09645299700000026

Fris, J., & Laridou, A. (2006). An additional way of thinking about organizational life and leadership: The quantum perspective. *Canadian Journal of Educational Administration and Policy*, 48.

Fullan, M. (2000). *Change forces: The sequel*. Philadelphia, PA: George H. Buchanan.

Gage, A. J. (2013). Association of child marriage with suicidal thoughts and attempts among adolescent girls in Ethiopia. *The Journal of Adolescent Health*, *52*(5), 654–656. doi:10.1016/j.jadohealth.2012.12.007 PMID:23433537

Garmston, R., & Bruce, W. (1995). Adaptive schools in a quantum universe. *Educational Leadership*, *52*(7), 6–12.

Gaurav, A., Mudit, K., et al. (2011). Distinction between formal and informal organisation. *Kalyan City Life*.

Genç, Z. (2010). *Web 2.0 yeniliklerinin eğitimde kullanımı: Bir facebook eğitim uygulama örneği. Akademik Bilişim'10 - XII*. Akademik Bilişim Konferansı Bildirileri 10 - 12 Şubat 2010 Muğla Üniversitesi.

Georgeta, I., & Castro, D. (2012). Governance in Spanish universities: Changing paradigms. *Journal of Hispanic Higher Education*, *11*(4), 336–350. doi:10.1177/1538192712441798

Ghaffar, A. (n.d.). Conflict in schools: Its causes & management strategies. *Journal of Management Sciences*, *11*(1), 213-229.

Gibson, J.L., Ivanvich, J., Donnelly, J., Jr., & Konopaske, R. (2006). *Organizations, behaviour & structure processes* (12th ed.). McGraw-Hill.

Gibson, J. L., & Ivancevich, J. M. et al.. (2006). *Organizations: Behaviour, structure, processes* (12th ed.). New York: McGraw-Hill, Irwin.

Gilbert, D. T., & Ebert, J. E. J. (2002). Decisions and revisions: The affective forecasting of changeable outcomes. *Journal of Personality and Social Psychology*, *82*(4), 503–514. doi:10.1037/0022-3514.82.4.503 PMID:11999920

Giles, C., & Hargreaves, A. (2006). The Sustainability of Innovative Schools as Learning Organizations and Professional Learning Communities During Standardized Reform. *Educational Administration Quarterly*, *42*(1), 124–156. doi:10.1177/0013161X05278189

Glanz, J., & Linda, S. B. (2000). *Paradigm debates in curriculum and supervision: Modern and Postmodern. Key Education Books*. Westport, CT: Bergin and Garvey.

Glasser, W. (1965). *Reality therapy: A new approach to psychiatry*. New York: Harper & Row.

Glasser, W. (1977). 10 steps to good discipline. *Today's Education*, *66*, 61–63.

Glasser, W. (1986). *Control therapy in the classroom*. New York: Harper & Row.

Glasser, W. (2009). *Classroom Management: Dealing with Discipline Problems*. Quality Education Programs, Inc. San Pedro.

Gleick, J. (1987). *Chaos: Making a new science.* New York: Penguin Books.

Gleitman, H., Fridlund, A. J., & Reisberg, D. (2000). *Social development. Basic Psychology* (pp. 480–517). New York: Norton.

Glickman, C. D. (1985). *Supervision of instruction: A developmental approach.* Boston: Allyn and Bacon.

Goldsmith, M., Greenberg, C., Robertson, A., & Hu-Chan, M. (2003). *Global leadership - The next generation.* Financial Times Press.

Goleman, D. (2013). İş başında duygusal zekâ (D. Balkara, H. Deniztekin, & O. Deniztekin, Trans.). İstanbul: Varlık Yayınları. (Original work published 1998).

Gomez, D. (2013). Leadership behavior and its impact on student's success and retention in online graduate education. *Academy of Educational Leadership Journal, 17*(2), 13–37.

Gordon, P. (1992). Perspectives and imperatives: Paradigms, transitions and the new supervision. *Journal of Curriculum and Instruction, 8*(1), 62–76.

Gordon, T. (1980). *Leadership effectiveness training- E. T.* New York: Bantam Books.

Gough, S., & Scott, W. (2008). *Higher education and sustainable development: Paradox and Possibility.* London: Routledge.

Grant, N. (2008). On the Usage of Social Networking Software Technologies in Distance Learning Education. In K. McFerrin et al. (Eds.), *Proceedings of Society for Information Technology and Teacher Education. International Conference.* Chesapeake, VA: AACE.

Grant, A. M., & Ashford, S. J. (2008). The dynamics of proactivity at work. *Research in Organizational Behavior, 28*, 3–34. doi:10.1016/j.riob.2008.04.002

Grant, A. M., & Parker, S. K. (2009). 7 Redesigning Work Design Theories: The Rise of Relational and Proactive Perspectives. *The Academy of Management Annals, 3*(1), 317–375. doi:10.1080/19416520903047327

Greaney, V., & Kallaghan, J. (1995). *Equity Issues in public examinations in developing countries.* World Bank Technical Paper No. 272.

Güngör, H. (2013). *Okul yönetici ve öğretmenlerinin damgalama eğilimleri ve örgüt kültürü arasındaki ilişki. Yayınlanmamış yüksek lisans tezi.* Sakarya Üniversitesi Eğitim Bilimleri Enstitüsü.

Güngör, H. (2014). Adolescence suicide as a chaotic phenomenon. In Ş. Ş. Erçetin & S. Banerjee (Eds.), *Chaos and Complexity in World Politics.* Hershey, PA: IGI Global.

Gunn, L. (1988). Public management: A third approach? *Public Money & Management, 8*(1-2), 21–25. doi:10.1080/09540968809387457

Gupton, S. L. (1998). *State-Mandated Reform of Programs Preparing School Principals: One University's Experience*. Paper presented at the Annual Meeting of the Mid-South Educational Research Association.

Hallinger, P. (2010). Developing Insructional Leadership. In Developing Successful Leadership, Studies in Educational Leadership 11. Springer Science+Business Media B.V.

Hambrick, D. C., Cho, T. S., & Chen, M. J. (1996). The influence of top management team heterogeneity on firms' competitive moves. *Administrative Science Quarterly*, *41*(4), 659–684. doi:10.2307/2393871

Hannah, S. T., Balthazard, P. A., Waldman, D. A., Jennings, P. L., & Tatcher, R. W. (2013). The psychological and neurological bases of leader self-complexity and effects on adaptive decision-making. *The Journal of Applied Psychology*, *98*(3), 393–411. doi:10.1037/a0032257 PMID:23544481

Hard, L., & Jónsdóttir, A. H. (2013). Leadership is not a dirty word: Exploring and embracing leadership in ECEC. *European Early*, *21*(3), 311–325. doi: 10.1080/1350293X.2013.814355

Hare, A. P. (1982). *Creativity in small groups*. Beverly Hills, CA: Sage.

Hassan, S., & Rohrbaugh, J. (2009). Incongruity in 360-degree feedback ratings and competing managerial values: Evidence from a public agency setting. *International Public Management Journal*, *12*(4), 421–449. doi:10.1080/10967490903103375

Heider, K. L. (2005). Teacher isolation: How mentoring programs can help. *Current Issues in Education, 8*(14). Available: http://cie.ed.asu.edu/volume8/number14/

Heilbrunn, S. (2010). Advancing entrepreneurship in an elementary school: A case study. *International Education Studies*, *3*(2), 174. doi:10.5539/ies.v3n2p174

Hendrick, R. (2003). Strategic planning environment, process, and performance in public agencies: A comparative study of departments in Milwaukee. *Journal of Public Administration: Research and Theory*, *13*(4), 491–519. doi:10.1093/jopart/mug031

Henning, J. E., & Robinson, V. (2002). Change and transformation inaneducational leadership program. In Changing World of School Administration. National Council of Professors of Educational Administration.

Hersey, P., & Blanchard, K. H. (1988). *Management of organizational behavior: Utilizing human resources* (5th ed.). Englewood Cliffs, NJ: Prentice Hall.

Hersey, P., Blanchard, K. H., & Dewey, E. J. (2002). *Management of organizational behaviour* (8th ed.). Prentice-Hall.

Holmes-Smith, P. (2006). *School socio-economic density and its effect on school performance. In SREAMS School Research Evaluation and Measurement Services. Australian-Sydney.* New South Wales Department of Education and Training.

Hooijberg, R., & Choi, J. (2001). The impact of organizational characteristics on leadership effectiveness models an examination of leadership in a private and a public sector organization. *Administration & Society*, *33*(4), 403–431. doi:10.1177/0095399701334002

Howard-Grenville, J. A. (2007). Developing issue-selling effectiveness over time: Issue selling as resourcing. *Organization Science*, *18*(4), 560–577. doi:10.1287/orsc.1070.0266

Howie, S. (2012). High-Stakes testing in South Africa: Friend or foe? *Assessment in Education: Principles, Policy & Practice*, *19*(1), 81–98. doi:10.1080/0969594X.2011.613369

Hoy, W. K., & Miskel, C. G. (2008). Educational administration: Theory, research and practice (8th ed.). New York. McGraw-Hill companies, Inc.

Iancu, D. (2012). Educational management, a new paradigm in social development.*International conference of scientific paper afases.*Brasov.

Ibara, E.C. (2008). Administering examinations for quality control in distance education: The National Open University of Nigeria perspective. *Turkish Online Journal ofDistance Education, 9*(3).

İli, K., & Erçetin, Ş. Ş. (2014). The impact of social media environments and tools on student behaviour. *US-China Education Review B, 4*(5), 295-302.

Informal Organization. (n.d.). In *Wikipedia, the free encyclopedia*. Retrieved on 4th November, 2014 from en.m.wikipedia.org/wiki/Informal_organization

Institute for Higher Education Policy. (2001). College and university Ranking system. *Global perspectives and American challenges*. Retrieved from www.ihep.org

Ishii, Y. (2003). *Development education in Japan: a comparative analysis of the contexts for its emergence, and its introdiction into the Japanese school system.* New York: Routledge Falmer.

Jamali, D. (2004). *Changing management paradigms: Implications for educational institutions. Suliman S.* Beirut, Lebanon: Olayan School of Business, American University of Beirut.

Jaspal, R. (2011). Caste, social stigma and identity processes. *Psychology and Developing Societies*, *23*(1), 27–62. doi:10.1177/097133361002300102

Johnson, D. W., & Johnson, R. T. (1987a). *Joining together: Group theory and group skills* (3rd ed.). Englewood Cliffs, NJ: Prentice Hall.

Johnson, D. W., & Johnson, R. T. (1987b). Research shows the benefits of adult cooperation. *Educational Leadership*, *45*(3), 27–30.

Johnson, D., Johnson, R., & Holubec, E. (1998). *Cooperation in the classroom.* Boston: Allyn and Bacon.

Jones, E., Farina, A., Hastorf, A., Markus, H., Miller, D., & Scott, R. (1984). *Social stigma: The psychology of marked to relationships*. New York: W.H. Freeman & Co.

Jones, F. (1979). The gentle art of classroom discipline. *National Elementary Principal*, *58*, 26–32.

Jones, P. W., & Coleman, D. (2005). *The United Nations and education. Multilateralism, development and globalisation*. New York, London: Routladge Falmer. doi:10.4324/9780203421499

Juenke, E. G. (2005). Management tenure and network time: How experience affects bureaucratic dynamics. *Journal of Public Administration: Research and Theory*, *15*(1), 113–131. doi:10.1093/jopart/mui006

Jullisson, EA, Karlsson, N, Gärling, T. (2005). Weighing the past and the future in decision making. *European Journal of Cognitive Psychology, 17*(4), 561-575. DOI: 10.1080/09541440440000159

Kara, A. (2008). *Understanding today's schools with Chaos Theory*. Academic Press.

Kara, Y. (2008). *Understanding today's schools with chaos theory. Fountain on life, knowledge and belief.* Retrieved from http://www.fountainmagazine.com/Issue/detail/Understanding-Todays-Schools-with-Chaos-Theory

Karadağ, E. (2010). Eğitim bilimleri doktora tezlerinde kullanılan araştırma modelleri: Nitelik düzeyleri ve analitik hata tipleri. *Kuram ve Uygulamada Eğitim Yönetimi*, *16*(1), 49–71.

Karaman, S., Yıldırım, S., & Kaban, A. (2008). *Öğrenme 2.0 yaygınlaşıyor: Web 2.0 uygulamalarının eğitimde kullanımına ilişkin araştırmalar ve sonuçları. XIII*. Ankara: Türkiye'de İnternet Konferansı Bildirileri, Orta Doğu Teknik Üniversitesi.

Karasar, N. (2006). *Bilimsel araştırma yöntemi*. Ankara: Nobel Yayın Dağıtım.

Karasar, N. (2010). *Bilimsel araştırma yöntemi (21. Baskı)*. Ankara: Nobel Yayın Dağıtım.

Kasl, E., & Yorks, L. (2002). Collaborative inquiry for adult learning. *New Directions for Adult and Continuing Education*, *94*(94), 3–11. doi:10.1002/ace.54

Kasliwal, P. (1995). *Development economics*. S. Western Collage Publishing.

Kasloff, P. (2007). *From teacher ısolation to professional collaboration: an experiment in school change*. Academic Press.

Kawesi, M. (2014). *School as formal organization*. Unpublished Manuscript, Department of Educational Management and Administration, Faculty of Education, Islamic University in Uganda.

Kayman, E. A. (2014). Chaos in education as an intelligent complex adaptive system. In Ş. Ş. Erçetin & S. Banerjee (Eds.), Chaos and Complexity Theory in World Politics (pp. 289-298). IGI Global.

Kayman, E. A. (2014). Chaos in education as an intelligent complex adaptive system. In Ş. Ş. Erçetin & S. Banerjee (Eds.), *Chaos and Complexity Theory in World Politics* (pp. 299–314). Global. doi:10.4018/978-1-4666-6070-0.ch019

Kayman, E. A., & Erçetin, Ş. Ş. (2011). The level of trained school principals' fulfillment of quantum leadership behaviors within Strengthening Vocational Education and Training (SVET) Project in Turkey. *Humanity and Social Sciences Journal*, *6*(1), 16–21.

Kaynak, M. (2014). Türkiye'de çocuk gelin sorunu. *Nesne, 2*(3), 27–38.

Kearns, L. (2011). High-stakes standardized testing & marginalized youth: An examination of the ımpact on those who fail. *Canadian Journal of Education, 34*(2), 112–130.

Keck, S. L., & Tushman, M. L. (1993). Environmental and organizational context and executive team structure. *Academy of Management Journal, 36*(6), 1314–1344. doi:10.2307/256813

Kellaghan, T., & Greaney, V. (2004). *Monitoring performance: Assessment and examinations in Africa.* Washington, DC: World Bank.

Kelly, K. F. (2001). *Meeting needs and making profits: The rise of for- profit degree- granting institutions. ECS Issue paper.* Denver, CO: Education Commission of the States.

Kimberly, J. R., & Evanisko, M. J. (1981). Organizational innovation: The influence of individual, organizational, and contextual factors on hospital adoption of technological and administrative innovations. *Academy of Management Journal, 24*(4), 689–713. doi:10.2307/256170 PMID:10253688

King, A. (2002). Structuring peer interaction to promote high-level cognitive processing. *Theory into Practice, 41*(1), 33–39. doi:10.1207/s15430421tip4101_6

Kirkpatrick, R., & Zang, Y. (2011). The negative ınfluences of exam-oriented education on chinese high school students: Backwash from classroom to child. *Language Testing in Asia, 1*(3), 3. doi:10.1186/2229-0443-1-3-36

Klar, H. W. & Brewer, C. A. (2013). Successful leadership in high-needs schools: an examination of core leadership practices enacted in challenging contexts. *Educational Administration Quarterly, 49*(5), 768–808. DOI: 10.1177/0013161X13482577

Kocabaşoğlu, N., & Aliustaoğlu, S. (2003). *Stigmatisation. New Symposium, 41*(4), 190–192.

Kocacık, E. ve Gökkaya, V. (2005). Türkiye'de çalışan kadınlar ve sorunları. *Cumhuriyet Üniversitesi İktisadi ve İdari Bilimler Dergisi, 6*(1), 195–219.

Koçak, K. (2000). Kaos ve Atmosfer. *TÜBİTAK Bilim ve Teknik, 391*, 94–97.

Kohn, A. (1996). *Beyond discipline: From compliance to community.* Alexandria, VA: Association for Supervision and Curriculum Development.

Kotter, J. P. (1999).*Leading change.* Cambridge, MA: Harvard Business School Press.

Kounin, J. S. (1970). *Discipline and group management in classrooms.* New York: Holt, Rinehart & Winston.

Krantz, D. H., & Kunreuther, H. C. (2007). Goals and plans in decision making. *Judgment and Decision Making, 2*(3), 137–168.

Larsen-Freeman, D. (1997). Chaos/complexityscience and second language acquisition. *Applied Linguistics, 18*(2), 141–165. doi:10.1093/applin/18.2.141

Leithwood, K., Seashore-Louis, K., & Leung, Y. F. (2010). *Conflict management and emotional intelligence*. (DBA thesis). Southern Cross University.

Leithwood, K., Louis, K. S., Anderson, S., & Wahlstrom, K. (2004). *How leadership influences student learning (paper commissioned by the Wallace Foundation)*. Minneapolis, MN: University of Minnesota.

Leithwood, K., Mascall, B., & Strauss, T. (2009). *Distributed Leadership According to the Evidence*. London: Rouhledge.

LePine, J. A., & Van Dyne, L. (1998). Predicting voice behavior in work groups. *The Journal of Applied Psychology*, *83*(6), 853–868. doi:10.1037/0021-9010.83.6.853

Lewin, K. M. (2003a). *Investigating the missing link—The case for expanded secondary schooling in Sub-Saharan Africa*. Academic Press.

Lewis, S., & Hardy, I. (2015, April3). Funding, reputation and targets: The discursive logics of high-stakes testing. *Cambridge Journal of Education*, *45*(2), 245–264. doi:10.1080/0305764X.2014.936826

Li, C. (2011). *Açık liderlik: Sosyal teknoloji yönetme tarzınızı nasıl değiştirir?* İstanbul: Optimist Yayınları.

Li, H., & Zhang, Y. (2007). The role of managers' political networking and functional experience in new venture performance: Evidence from China's transition economy. *Strategic Management Journal*, *28*(8), 791–804. doi:10.1002/smj.605

Link, B. G., & Phelan, J. C. (2001). Conceptualizing stigma. *Annual Review of Sociology*, *27*(1), 363–385. doi:10.1146/annurev.soc.27.1.363

London, M., Larsen, H. H., & Thisted, L. N. (1999). Relationships between feedback and self-development. *Group & Organization Management*, *24*(1), 5–27. doi:10.1177/1059601199241002

Lorenz, E. N. (1995). *The essence of chaos*. University of Washington Press.

Lortie, D. C. (1975). *School teacher*. Chicago: University of Chicago Press.

Lucia, A. (1997). Leaders know how to listen. *HRFocus*, *74*(4), 25.

Luo, Y. (2003). Industrial dynamics and managerial networking in an emerging market: The case of China. *Strategic Management Journal*, *24*(13), 1315–1327. doi:10.1002/smj.363

Madaus, G., & Clarke, M. (2001). The adverse impact of high-stakes testing on minority students: Evidence from one hundred years of test data. In G. Orfield & M. L. Kornhaber (Eds.), *Raising standards or raising barriers? Inequality and high-stakes testing in public education*. New York: The Century Foundation Press.

Maddock, R. C., & Fulton, R. L. (1998). *Motivation, emotions, and leadership: The silent side of management*. Greenwood Publishing Group.

Major, B., & O'Brien, L. T. (2005). The social psychology of stigma. *Annual Review of Psychology, 56*(1), 393–421. doi:10.1146/annurev.psych.56.091103.070137 PMID:15709941

Major, D. A., & Kozlowski, S. W. (1997). Newcomer information seeking: Individual and contextual influences. *International Journal of Selection and Assessment, 5*(1), 16–28. doi:10.1111/1468-2389.00042

Mandelbrot, B. B. (1983). *The fractal geometry of nature.* W. H. Freeman and Company.

Manga, S. D. (2010). *Administration of higher education. Administration and planning.* Unpublished manuscript, Usmanu Danfodiyo University, Sokoto Nigeria.

Maranto, R. (2003). Lobbying in Disguise. *Education Next, 3*(1), 79–84.

Marion, R. (2006). Complexity in organizations: a paradigm shift. In A. Sengupta (Ed.), *Chaos, Nonlinearity, Complexity* (pp. 247–269). Springer Berlin Heidelberg. doi:10.1007/3-540-31757-0_9

Marton, F., & Säljö, R. (1984). Approaches to learning. In F. Marton, D. Hounsell, & N. Entwistle (Eds.), *The Experience of Learning.* Edinburgh, UK: Scottish Academic Press.

Marton, F., & Saljo, R. (1997). Approaches to learning. In F. Marton, D. Hounsell, & N. Entwistle (Eds.), *The experience of learning* (pp. 3–22). Edinburgh, UK: Scottish Academic Press.

Marzano, R., Waters, T., & McNulty, B. (2005). *School leadership that works: from research to results.* Aurora, CO: Mid-Continent Research for Education and Learning.

Mason, M. (2008). Complexity theory and the philosophy of education. *Educational Philosophy and Theory, 40*(1), 4–18. doi:10.1111/j.1469-5812.2007.00412.x

Massarik, F., & Pei-Carpenter, M. (2002). *Organization development and consulting.* San Francisco: Jossey- Bass/Pfeiffer.

Mathew, M. (2014). Introducing effective problem solving culture in higher education institutions in Nigeria: Methods and tools. *Journal of Education and Practice, 5*(25).

Mayer, J. D., & Salovey, P. (1990). Emotional intelligent. Bywood Publishing Co., Inc.

Mayer, R. E., & Wittrock, R. C. (2006). Problem solving. In P. A. Alexander & P. H. Winne (Eds.), *Handbook of educational psychology* (2nd ed.; pp. 287–304). Mahwah, NJ: Erlbaum.

McCallum, S. Y. L., Forret, M., & Wolff, H. G. (2014). Internal and external networking behavior: An investigation of relationships with affective, continuance, and normative commitment. *Career Development International, 19*(5), 595–614. doi:10.1108/CDI-08-2013-0101

McCarthy, M. M. (2002). The changing environment for school leaders: market forces. In Changing World of School Administration. National Council of Professors of Educational Administration.

McEwan, E. (2003). *10 Traits of Highly Effective Principals: From Good To Great Performance.* Thousand Oaks, CA: Corwin Press, Inc.

Mcloughlin, C., & Lee, M. J. (2007). Social software and participatory learning: Pedagogical choices with technology affordances in the Web 2.0 Era. In ICT: Providing Choices For Learners And Learning. Proceedings Ascilite Singapore.

McMillan, E. (2004). *Complexity, Organizations and Change*. London: Routledge, Taylor and Francis Group.

McNeil, J. D. (1996). *Curriculum: A comprehensive introduction*. New York, NY: Harper-Collins Publishers.

McPherson, (1970). *Small town teacher. In sociological work*. Chicago: Aldine Publishing Company.

Menteşe, M. (2013). *Sosyal medya ortam ve araçlarının eğitimde kullanımına ilişkin okul yöneticilerinin ve öğretmenlerin görüşleri*. Ankara: Yüksek Lisans Tezi, Hacettepe Üniversitesi.

Miller, D., Lant, T. K., Milliken, F. J., & Kom, H. J. (1996). The evolution of strategic simplicity: Exploring two models of organizational adaption. *Journal of Management*, *22*(6), 863–887. doi:10.1177/014920639602200604

Miller, J. P. (1994). The relationship between organizational culture and environmental scanning: A case study. *Library Trends*, *43*(2), 170–205.

Miller, V. D., & Jablin, F. M. (1991). Information seeking during organizational entry: Influences, tactics, and a model of the process. *Academy of Management Review*, *16*(1), 92–120.

Mirel, J. & Goldin, S. (2012). *Alone in the classroom: Why teachers are too isolated*. Academic Press.

Mohamed, A., & Abdelraheem. (2005). *Problem solving and decision making published by center of advancement of postgraduate studies and research in engineering sciences*. Faculty of Engineering, Cairo University (CAPSCU).

Morgan, G. (1998). *Images of organization*. SAGE Publications.

Morgan, H. (2003). Challenges of a global marketplace. In Human Resources in the 21st Century. San Francisco, CA: John Wiley &. Sons.

Morrison, K. (2006). *Complexity Theory and education*. APERA Conference 2006, Hong Kong.

Morrison, E. W. (1993). Longitudinal study of the effects of information seeking on newcomer socialization. *The Journal of Applied Psychology*, *78*(2), 173–183. doi:10.1037/0021-9010.78.2.173

Morrison, E. W., & Phelps, C. C. (1999). Taking charge at work: Extrarole efforts to initiate workplace change. *Academy of Management Journal*, *42*(4), 403–419. doi:10.2307/257011

Morrison, K. (2002). *School Leadership and Complexity Theory*. New York: Routledge.

Morrison, K. (2008). Educational Philosophy and the Challenge of Complexity Theory. *Educational Philosophy and Theory*, *40*(1), 19–34. doi:10.1111/j.1469-5812.2007.00394.x

Morrison, K. (2015). Complexity theory, school leadership and management: Questions for theory and practice. *Educational Management Administration & Leadership*, *38*(3), 374–393. doi:10.1177/1741143209359711

Msila, V. (2011). School management and the struggle for effective schools. *Africa Education Review*, *8*(3), 434–451. doi:10.1080/18146627.2011.618650

Muijs, D., West, M., & Ainscow, M. (2010). Why network? Theoretical perspectives on networking. *School Effectiveness and School Improvement*, *21*(1), 5–26. doi:10.1080/09243450903569692

Mulford, B. (2003). *School leaders: Changing roles and impact on teacher and school effectiveness.* A paper commissioned by the Education and Training Policy Division, OECD, for the Activity Attracting, Developing and Retaining Effective Teachers.

Mullins, L. J. (2010). Management & organizational behavior (9th ed.). Pearson Prentice Hall

Munoz, C., & Towner, T. (2009). Opening facebook: How to use facebook in the college classroom. In Society For İnformation Technology & Teacher Education İnternational Conference.

Murphy, J., Elliot, S., Goldring, E., & Porter, A. (2006). *Learning-Centered Leadership: A Conceptual Foundation.* Paper presented at the annual meeting of the Wallace Foundation State Action for Educational Leadership Conference, Saint Louis, MO.

Murphy, J., Elliot, S., Goldring, E., & Porter, A. (2007). Leadership for learning: a research-based model and taxonomy of behaviors. *Journal of School Leadership and Management, 27*(2), 179–201. Retrieved November, 3 2014, from http://www.valed.com/documents/3_murphy%20et%20al_2007.pdf

Murphy, J., & Hallinger, P. (1988, February). The characteristics of instructionally effective school districts. *The Journal of Educational Research*, *81*(3), 176–181. doi:10.1080/00220671.1988.10885819

Murray, J. (2007). Countering insularity in teacher education: academic work on pre-service courses in nursing, social work and teacher education. *Journal of Education for Teaching: International Research and Pedagogy, 33*, 3.

Murray, C. (2008). Schools and social networking: Fear or education? *Synergy Perspectives: Local*, *6*(1), 8–12.

Musisi, B. (2012). *Pedagogic Ramifications of the competitive Secondary Education market in Uganda: A case of Proprietary schools.* (Unpublished PhD Thesis). School of Education, Makerere University, Kampala.

Musoleno, R. R., & White, G. P. (2010). Influences of high-stakes testing on middle school mission and practice. RMLE Online. *Research in Middle Level Education*, *34*(3), 1–10.

Nastase, M. (2007). Types of organisational csonflicts. *Revista de Management Comparat Internaţional*, *8*(4), 80.

Nawait, M.S., Sirajo, A., & Nasir, Z.M. (2009). Multi-Disciplinary conference. *Farfaru Journal, 4*.

Neeraj, K. (2006). *Principles of management*. Jaipur, India: Sublime Publications.

Newell, A., & Simon, H. A. (1972). *Human problem solving*. Englewood Cliffs, NJ: Prentice-Hall.

Newman, J., Raine, J., & Skelcher, C. (2000). *Innovation and best practice in local government: A research report*. Department of the Environment, Transport and the Regions.

Nichols, S. L., Glass, G. V., & Berliner, D. C. (2012). High-stakes testing and student achievement: *Updated Analyses with NAEP Data. Education Policy Analysis Archives, 20*(20), 35.

Nweze, T. (2009). Management of examinations: Ethical issues. *Edo Journal of Counselling, 2*, 1.

Ogawa, R. T., & Bossert, S. T. (1995). Leadership as an organizational quality. *Educational Administration Quarterly, 31*(2), 224–243. doi:10.1177/0013161X95031002004

Ogbeifum, S. & Olisa,E. (2001, July 1). Half-baked teachers bane of education woes. *The Vanguard Online*.

Oladeji, B. (2001, August 2). More private varsities coming FG. *The Nigerian Tribune*.

O'Neil, J. (2004). Effects of gender and power on PR managers' upward influence. *Journal of Managerial Issues*, 127–144.

Onuorah, M., & Aniete, B. A. (2001). Govt Releases N1b To Pay Police. *The Guardian*.

Oran, N. T., & Şenuzun, F. (2008). Toplumda kırılması gereken bir zincir: HIV/AIDS stigması ve baş etme stratejileri. *Uluslararası İnsan Bilimleri Dergisi, 5*(1), 1–16.

Othman, R., & Hamedon, S. R. (2011). *Environmental scanning practice of enterprise 50 Small Medium Enterprise (E50 SMEs) in Malaysia*. Retrieved from irep.iium.edu.my

Özbaş, M. (2010). İlköğretim okullarında öğrenci devamsızlığının nedenleri – The reasons of absenteeism in primary schools. *Eğitim ve Bilim – Education and Science, 35*(156), 32-44.

Özbaş, M. (2014). İlköğretim okulu yöneticilerinin aile memnuniyetini karşılama düzeyine ilişkin veli görüşleri. *Eğitim Bilimleri Araştırmaları Dergisi- Uluslararasi E-Dergi-Journal of Educational Sciences Research-International E-Journal. EBAD-JESR, 4*(1), 243-258. 10.12973/jesr.2014.41.12".2014.41.12

Özbaş, M., & Avcı, M. (2013). *Yoksul aile çocuklarının kamusal eğitimden yararlanma düzeylerine ilişkin veli algıları*. IASSR European Conference on Social Science Research. Marmara University, School of Business Administration, İstanbul.

Özbaş, M. (2009). *İlköğretim okulu yöneticilerinin okul-aile ilişkileri konusunda yapmaları gereken ve yapmakta oldukları işler. In Yayımlanmamış Doktora Tezi*. Ankara: Hacettepe Üniversitesi Sosyal Bilimler Enstitüsü.

Özbaş, M. (2012). Kız çocuklarının ortaöğretimde okullaşma oranlarına etki eden nedenlere ilişkin algıları - Perceptions of girls related to the reasons that affect schooling rates in secondary education. *IJTASE International Journal of New Trends in Arts, Sports & Science Education, 1*(4), 60–71.

Özdemir, S., & Kılınç, A. Ç. (2014). Değişen toplum ve okul. In. S. Özdemir (Ed.), Türk eğitim sistemi ve okul yönetimi (pp. 75-91). Ankara: Pegem Akademi.

Özkan, B., & Mckenzie, B. (2008). Social networking tools for teacher education. In K. Mcferrin et al. (Eds.), *Proceedings of Society for Information Technology and Teacher Education International Conference*. Chesapeake, VA: AACE.

Özmen, F., Aküzüm, C., Sünkür, M., & Baysal, N. (2011). Sosyal ağ sitelerinin eğitsel ortamlardaki işlevselliği. In *6th International Advanced Technologies Symposium (IATS'11)*.

Öztürk, M., & Akgün, Ö. E. (2012). Üniversite öğrencilerinin sosyal paylaşım sitelerini kullanma amaçları ve bu sitelerin eğitimlerinde kullanılması ile ilgili görüşleri. *Sakarya University Journal of Education, 2*(3), 5–27.

Parker, R., & Aggleton, P. (2003). HIV and AIDS-related stigma and discrimination: A conceptual framework and implications for action. *Social Science & Medicine, 57*(1), 13–24. doi:10.1016/S0277-9536(02)00304-0 PMID:12753813

Parker, S. K. (1998). Enhancing role breadth self-efficacy: The roles of job enrichment and other organizational interventions. *The Journal of Applied Psychology, 83*(6), 835–852. doi:10.1037/0021-9010.83.6.835 PMID:9885197

Parker, S. K., & Collins, C. G. (2010). Taking stock: Integrating and differentiating multiple proactive behaviors. *Journal of Management, 36*(3), 633–662. doi:10.1177/0149206308321554

Pelled, L. H. (1996). Demographic diversity, conflict, and work group outcomes: An intervening process theory. *Organization Science, 7*(6), 615–631. doi:10.1287/orsc.7.6.615

Perry, J. L., & Rainey, H. G. (1988). The public-private distinction in organization theory: A critique and research strategy. *Academy of Management Review, 13*(2), 182–201.

Peter, A., et al. (2011). *Challenges for faculty managements at African higher education institutions*. Germany University of Applied Science, Osnabruck.

Poluku, K. (2013). Analysis of conflict management and leadership for organizational change. *International journal of Research in social science*. Retrieved from www.ijsk.org/ijras

Preble, J. F. (1992). Towards a Comprehensive System Of Strategic Control. *Journal of Management Studies, 29*(4), 391–408. doi:10.1111/j.1467-6486.1992.tb00671.x

Priye, S. T. (2011). *The corrosive effect of corruption on Nigerian educational system*. Academic Press.

Quisumbing, L. R. (2001, January 10). *Changing paradigms for teacher education for the new millennium*. Paper read at the International Conference on Teacher Education, Philippine Normal University of Manila.

Radford, M. (2006). Researching classrooms: Complexity and chaos. *British Educational Research Journal, 32*(2), 177–190. doi:10.1080/01411920600568950

Rainey, H. G. (1983). Public Agencies and Private Firms Incentive Structures, Goals, and Individual Roles. *Administration & Society, 15*(2), 207–242. doi:10.1177/009539978301500203

Raja, A. S. & Palanichamy, P. (2011). Leadership styles of school principals- an examination at Madurai district. *Global Management Review, 5*(4).

Rashman, L., Withers, E., & Hartley, J. (2009). Organizational learning and knowledge in public service organizations: A systematic review of the literature. *International Journal of Management Reviews, 11*(4), 463–494. doi:10.1111/j.1468-2370.2009.00257.x

Rausch, E., Halfhill, S. M., Sherman, H., & Washbush, J. B. (2001). Practical leadership-in-management education for effective strategies in a rapidly changing world. *Journal of Management Development, 20*(3), 245–257. doi:10.1108/02621710110386381

Reddell, S. (2010). *High stakes testing: Our children at risk.* Online Submission.

Reichers, A. E. (1987). An Interactionist Perspective on Newcomer Socialization Rates. *Academy of Management Review, 12*(2), 278–287.

Reid, W. A., Duvall, E., & Evans, P. (2007). Relationship between assessment results and approaches to learning in year two medical students. *Medical Education, 41*(8), 754–762. doi:10.1111/j.1365-2923.2007.02801.x PMID:17661883

Renn, R. W., & Fedor, D. B. (2001). Development and field test of a feedback seeking, self-efficacy, and goal setting model of work performance. *Journal of Management, 27*(5), 563–583. doi:10.1177/014920630102700504

Reuters. (2001). *Nigerian says killed minister for one min naira*. Ibadan, Nigeria: Reuters.

Richardson, J. T. E. (2003). Approaches to studying and perceptions of academic quality in a short web-based course. *British Journal of Educational Technology, 34*(4), 433–442. doi:10.1111/1467-8535.00340

Ritz, J. M., Deal, W. F., Hadley, F., Jacobs, J. A., Kildruff, T. F., & Skena, K. G. (1986a). Problem-solving. *Technology Teacher, 46*(2), 15–22.

Robinson, K. (2008). *Changing education paradigms*. RSA.

Robinson, V., Lloyd, C., & Rowe, K. (2008). The impact of leadership on student outcomes: An analysis of the differential effects of leadership types. *Educational Administration Quarterly, 44*(5), 635–674. doi:10.1177/0013161X08321509

Rosado, C. (1996, December 15). *Restructuring education for the 21st century*. Department of Urban Studies, Eastern University.

Rousseau, J. J. (1994). *The Social Contract*. Oxford University Press.

Routledge. (2013). *Global corruption report: Education transparency international*. Author.

Russell, R. D., & Russell, C. J. (1992). An Examination of the Effects of Organizational Norms, Organizational Structure, Environmental Uncertainty on Entrepreneurial Strategy. *Journal of Management*, *18*(4), 639–656. doi:10.1177/014920639201800403

Sagi, A., & Friedland, N. (2007). The cost of richness: The effect of the size and diversity of decision sets on post-decision regret. *Journal of Personality and Social Psychology*, *93*(4), 515–524. doi:10.1037/0022-3514.93.4.515 PMID:17892329

Salanova, M., & Schaufeli, W. B. (2008). A cross-national study of work engagement as a mediator between job resources and proactive behaviour. *International Journal of Human Resource Management*, *19*(1), 116–131. doi:10.1080/09585190701763982

Saleemi, N. A. (2011). *Principles and practice of management simplified* (Revised Edition). Nairobi, Kenya: Saleemi Publications Limited.

Salusu, J. (2000). Pengambilan keputusan stratejikuntuk organisasi publik dan organisasi non profit. Jakarta: Gramedia Widiasarana Indonesia.

Salvatore, P., Indic, P., Murray, G., & Baldessarini, R. J. (2012). Article. *Dialogues in Clinical Neuroscience*, *14*(4), 369–379. PMID:23393414

Sandra, J. (2013). On the reliability of high-stakes teacher assessment. *Research Papers in Education*, *28*, 1.

Saphier, J., & Glower, R. (1997). *The skillful teacher: Building your teaching skills*. Research for Better Teaching, Inc.

Sardar, Z., & Abrams, I. (1999). *Introducing Chaos*. Icon Books Ltd.

Sass, C. P. (1994). *On Interpersonal Competence. Interpretive Approaches to Interpersonal Communication* (K. Carter & M. Presnell, Eds.). Albany, NY: State University of New York Press.

Schein, E. H. (1992). *Organisational culture and leadership* (2nd ed.). San Francisco: Jossey-Bass.

Schlechty, P. C. (2011). *Okulu yeniden kurmak* (Y. Özden, Trans.). Ankara: Nobel Yayıncılık.

Scott, S. G., & Bruce, R. A. (1994). Determinants of innovative behavior: A path model of individual innovation in the workplace. *Academy of Management Journal*, *37*(3), 580–607. doi:10.2307/256701

Şeker, M. (2011). *İstanbul'da Yaşam Kalitesi Araştırması*. İstanbul: İstanbul Ticaret Odası yayınları.

Sengupta, N. (2009). *Emotional intelligence*. New Delhi: Excel Books.

Seyyedrezae, H. S. (2014). The application of chaos/complexity theory in classroom teaching, task design and lesson planning. *Journal of Language Sciences & Linguistics*, *2*(2), 27–32.

Sharma, M. (2003). *Theory of public administration*. New Delhi, India: Anmol Publications Private Limited..

Shelton, M. M., & Bauer, L. K. (1994). *Secrets of highly effective meetings*. Thousand Oaks, CA: Corwin Press.

Shyamala, P. (n.d.). Leadership during times of complexity and changes: Human resource area. *Journal of Business and Management*.

Siagian, S. P. (1987). *Filsafat Administrasi*. Jakarta: PT Gunung Agung.

Sieben, B., & Wettergren, A. (Eds.). (2010). *Emotionalizing organizations and organizing emotions*. Palgrave Macmillan. doi:10.1057/9780230289895

Sığrı, U. (2010). The paradigm shift in educational management: An evaluation of distributed learning as future approach. *The Turkish Online Journal of Educational Technology*, *9*(2), 137–149.

Şimşek, Y. (2003). *Okul müdürlerinin iletişim becerileri ile örgüt kültürü arasındaki ilişki-Eskişehir ili örneği. Yayımlanmamış doktora tezi*. Eskişehir: Anadolu Üniversitesi, Eğitim Bilimleri Enstitüsü.

Şişman, M. (1994). *Örgüt Kültürü: Eskişehir il merkezindeki ilkokullarda bir araştırma*. Anadolu Üniversitesi Eğitim Fakültesi Yayınları.

Škerlavaj, M., Štemberger, M. I., & Dimovski, V. (2007). Organizational learning culture—the missing link between business process change and organizational performance. *International Journal of Production Economics*, *106*(2), 346–367. doi:10.1016/j.ijpe.2006.07.009

Skinner, B. F. (1968). *The technology of teaching*. New York: Appleton-Century-Crofts.

Skinner, B. F. (1971). *Beyond freedom and dignity*. New York: Knopf.

Skule, S. (2004). Learning conditions at work: A framework to understand and assess informal learning in the workplace. *International Journal of Training and Development*, *8*(1), 8–20. doi:10.1111/j.1360-3736.2004.00192.x

Slavin, R. E. (1987). Cooperative learning and the co- operative schools. *Educational Leadership*, *45*(3), 7–13.

Snyder, K. J., Acker-Hocevar, M., & Wolf, K. M. (1995). *Chaos Theory as a Lens for Advancing Quality Schooling*. Academic Press.

Sonfield, M., Lussier, R., Corman, J., & McKinney, M. (2001). Gender comparisons in strategic decision-making: An empirical analysis of the entrepreneurial strategy matrix. *Journal of Small Business Management*, *39*(2), 165–173. doi:10.1111/1540-627X.00015

Sousa, D. A. (2001). How the brain learns (2nd ed.). Academic Press.

Steyn, G. M. (2013). Building professional learning communities to enhance continuing professional development in South African Schools. *Anthropologist, 15*(3), 277–289.

Stigma ve Epilepsi . (2013). Retrieved from www.turkepilepsi.org.tr/userfiles/file/.../stigma_ve_epilepsi.pdf

Stogdill, R. M. (1948). Personal factors associated with leadership: A survey of the literature. *The Journal of Psychology, 25*(1), 35–71. doi:10.1080/00223980.1948.9917362 PMID:18901913

Stoner, J. L. (2012). *Manage the challenges of working ın a matrix organization.* Retrieved on 25th mar, 2015 from seapointcenter.com/matrix-leadership/

Syamsi, I. (2000). *Pengambilan keputusan dan sistem ınformasi.* Jakarta: Bumi Aksara.

Taner, Y., & Gökler, B. (2004). Çocuk istismar ve ihmali: Psikiyatrik yönleri. *Hacettepe Tıp Dergisi, 35*, 82–86.

Taş, U., & Yenilmez, F. (2008). Türkiye'de eğitimin kalkınma üzerindeki rolü ve eğitim yatırımlarının geri dönüş oranı. *Eskişehir Osmangazi Üniversitesi Sosyal Bilimler Dergisi, 9*(1), 155–186.

TEDMEM. (2015). *Siyasi partilerin 2015 genel seçim bildirgeleri: Eğitim politikalarına ilişkin değerlendirmeler. In Güncel Yayınlar G3.* Ankara: Türk Eğitim Derneği TEDMEM.

Tetenbaum, T. J. (1998). Shifting paradigms: From Newton to chaos. *Organizational Dynamics, 26*(4), 21–32. doi:10.1016/S0090-2616(98)90003-1

Teunissen, P. W., Stapel, D. A., van Der Vleuten, C., Scherpbier, A., Boor, K., & Scheele, F. (2009). Who wants feedback? An investigation of the variables influencing residents' feedback-seeking behavior in relation to night shifts. *Academic Medicine, 84*(7), 910–917. doi:10.1097/ACM.0b013e3181a858ad PMID:19550188

The Davies Report. (1994). *The 'great battle' in Swansea. Times higher education supplement.* Bristol, UK: Thoemmes Press.

The Guardian Online. (2001, August 2). *Money and politics of bitterness.* Author.

The Need for Paradigms and Models: Excerpted from Samuel Huntington. (1996). The New Era in World Politics. In *The Clash of Civilizations and the Remaking of World Order.* Retrieved from http://www.washingtonpost.com/wp-srv/style/longterm/books/chap1/clashofcivilizations.htm

The World Bank. (2006). *World Development Report 2007: Development and the Next Generation.* The International Bank for Reconstruction and Development The World Bank.. DOI: 10.1596/978-0-8213-6549-6

Thomas, J. B., Clark, S. M., & Gioia, D. A. (1993). Strategic sensemaking and organizational performance: Linkages among scanning, interpretation, action, and outcomes. *Academy of Management Journal, 36*(2), 239–270. doi:10.2307/256522 PMID:10125120

Thomas, J. B., Sussman, S. W., & Henderson, J. C. (2001). Understanding "strategic learning": Linking organizational learning, knowledge management, and sensemaking. *Organization Science, 12*(3), 331–345. doi:10.1287/orsc.12.3.331.10105

Thompson, J. A. (2005). Proactive personality and job performance: A social capital perspective. *The Journal of Applied Psychology, 90*(5), 1011–1017. doi:10.1037/0021-9010.90.5.1011 PMID:16162073

Thompson, S. C., Gregg, L., & Niska, J. M. (2004). Professional Learning Communities, Leadership, and Student Learning. *RMLE Online: Research in Middle Level Education, 28*(1), 1–15. doi:10.1080/19404476.2004.11658173

Tillman, L. C. (2002). The impact of diversity in educational administration. In Changing World of School Administration. National Council of Professors of Educational Administration.

Timberlake, G. R. (2004). Decision-making in multi-campus higher education institutions. *Community College Enterprise, 10*(2), 91-99.

Tiryakioğlu, F., & Erzurum, F. (2011). Bir eğitim aracı olarak ağların kullanımı. In *2nd International Conference on New Trends in Education and Their Implications*. Retrieved from www.iconte.org

Titcomb, T. J. (1998). *Chaos and complexity theory (No. 9807)*. American Society for Training and Development.

Tobin, J. (2014). Management and leadership issues for school building leaders. *International Journal of Educational Leadership Preparation, 9*(1).

Todd, S. Y., Harris, K. J., Harris, R. B., & Wheeler, A. R. (2009). Career success implications of political skill. *The Journal of Social Psychology, 149*(3), 279–304. doi:10.3200/SOCP.149.3.279-304 PMID:19537596

Toğay, A., Akdur, T. E., Yetişen, İ. C., & Bilici, A. (2013). *Eğitim süreçlerinde sosyal ağların kullanımı bir MYO deneyimi*. XV. Akademik Bilişim Konferansı, Akdeniz Üniversitesi, Antalya. Retrieved from http://ab.org.tr/ab13/bildiri/302.pdf

Tripati, P. C., & Reddy, P. N. (2008). *Principles of management (4th ed.)*. New Delhi, India: Tata McGraw-Hill Education Private Limited.

Trygestad, J. (1997). *Chaos in the classroom: An application of chaos theory*. Roundtable Session of Chaos and Complexity Theory. American Educational Research Association. University of Minnesota. Retrieved from www.eric.ed.gov

Tugba, A. (2012). *An exploration of the effects of high-stakes examinations on the teaching and learning of mathematics in post-primary education in Ireland and Turkey*. (PhD thesis). National University of Ireland, Maynooth, Ireland.

Tuncer, A. M., & Açıkalın, Ş. N. (2015). Solution to chaotic situations in higher education: New generation universities as intelligent organizations. In Ş.Ş. Erçetin & S. Banerjee (Eds.), Chaos, Complexity and Leadership 2013 (pp. 1-9). Springer International Publishing.

U S National Library of Medicine National Institute of Health. (2007). A simple guide to chaos and complex. *Journey of Epidemiology and Community Health, 61*(11), 933-937.

UBEC. (2012). *About UBEC*. Universal Basic Education Commission.

UBEC. (2012). About UBEC. Universal Basic Education Commission. Retrieved 30 August 2012.

Üçok, A. (2003). Şizofreni hastası neden damgalanır? *Klinik Psikiyatri, 1*, 3–8.

Uğurlu, Z. (2015). Okul örgütü ve yönetimi. In *Türk eğitim sistemi ve okul yönetimi* (pp. 181–237). Ankara: Pegem Akademi.

Uhl-Bien, M., Marion, R., & McKelvey, B. (2007). *Complexity Leadership Theory: Shifting Leadership from the Industrial Age to the Knowledge Era*. Lincoln, NE: Leadership Institute Faculty Publications.

Umar, B. & Adoba I. (2001, December 6). Senate Passes UBE Bill. *ThisDay*.

UNAIDS. (2000). HIV/AIDS and the Education Sector. UNAIDS.

UNESCO. (2010). *Why gender equality in basic education in Pakistan? United Nations Educational, Scientific and Cultural Organization*. Islamabad: UNESCO Office, Serena Business Complex.

UNESCO. (2011). *World data on Education*. UNESCO-IBE.

Ural, S. S., & Açıkalın, Ş., N. (2015). Protection of fundamental rights and freedoms as a universal problem and chaos theory. In Ş.Ş. Erçetin & S. Banerjee (Eds.), Chaos, Complexity and Leadership 2013 (pp. 197-202). Springer International Publishing.

VandeWalle, D., Ganesan, S., Challagalla, G. N., & Brown, S. P. (2000). An integrated model of feedback-seeking behavior: Disposition, context, and cognition. *The Journal of Applied Psychology, 85*(6), 996–1003. doi:10.1037/0021-9010.85.6.996 PMID:11125662

Varas-Díaz, N., Neilands, T. B., Malavé Rivera, S., & Betancourt, E. (2010). Religion and HIV/AIDS stigma: Implications for health professionals in Puerto Rico. *Global Public Health: An International Journal for Research, Policy and Practice*, 1–18. Retrieved from http://www.ncbi.nlm.nih.gov/pmc/articles/PMC2891884/ PMID:20087809

Viacava, K. R., & Pedrozo, E. A. (2010). Higher education in management: Reinventing the paradigm to gain the capacity to handle today's complexity. *On the Horizon, 18*(1), 45–52. doi:10.1108/10748121011020994

Vocational Education in Nigeria. (2012). UNESCO-UNEVOC.

Wahlstrom, K. L., Louis, K. S., Leithwood, K., & Anderson, S. E. (2010). *Learning from Leadership: Investigating the Links to Improved Student Learning*. St. Paul, MN: University of Minnesota.

Wallace-Bruce, N. L. A. (2010). *Assessment of factors affecting female participation in senior high school education in Ghana: A case study of Asunafo North and Sunyani Municipalities and Kumasi metropolitan area. Master of Science the School of Graduate Studies.* Ghana: Kwame Nkrumah University of Science and Technology, in Development Policy and Planning Department of Planning College of Architecture and Planning.

Wanberg, C. R., & Kammeyer-Mueller, J. D. (2000). Predictors and outcomes of proactivity in the socialization process. *The Journal of Applied Psychology*, *85*(3), 373–385. doi:10.1037/0021-9010.85.3.373 PMID:10900812

Wang, D. B. (2004). Family background factors and mathematics success: A comparison of Chinese and US students. *Educational Research*, *41*, 40–54.

Wanger. (2008). *The global achievement gap: Why even our best schools don't teach the new survival skills our children need---and what we can do about it.* Academic Press.

Wang, J., Tolson, H., Chiang, T. L., & Huang, T. Y. (2010). An exploratory factor analysis of workplace learning, job satisfaction, and organizational commitment in small to mid-size enterprises in Taiwan. *Human Resource Development International*, *13*(2), 147–163. doi:10.1080/13678861003703682

Ward, C. J. (2013). Why leadership matters: one school's journey to success. *Educational Leadership and Administration: Teaching and Program Development, 24*, 62-73.

Warren, H. (2004). *Deep and surface approaches to learning: Learning and teaching theory for engineering academics. engineering subject centre guide*. Loughborough: HEA.

Wasley, P. (2012). *The detriment of teacher ısolation*. Academic Press.

Weng, C.-H., & Tang, Y. (2014). The relationship between technology leadership strategies and effectiveness of school administration: An empirical study. *Computers & Education*, *76*, 91–107. doi:10.1016/j.compedu.2014.03.010

Wester, S. R., Christianson, H. F., Fouad, N. A., & Santiago-Rivera, A. L. (2008). Information processing as problem solving: A collaborative approach to dealing with students exhibiting insufficient competence. *Training and Education in Professional Psychology*, *2*(4), 193–201. doi:10.1037/1931-3918.2.4.193

West, M. A., & Anderson, N. R. (1996). Innovation in top management teams. *The Journal of Applied Psychology*, *81*(6), 680–693. doi:10.1037/0021-9010.81.6.680

Wheatley, M. (2003). When change is out of our control. In Human Resources in the 21st Century. San Francisco: John Wiley & Sons.

Wheatley, M. J. (2010). *Leadership and the new science: Discovering order in a chaotic world.* San Francisco: Berrett-Koehler Publishers.

Wood, S. L. (2001). Remote purchase environments: The influence of return policy leniency on two-stage decision process. *JMR, Journal of Marketing Research, 38*(2), 157–169. doi:10.1509/jmkr.38.2.157.18847

Woods, D. (1985). Problem-based learning and problem solving. In D. Boud (Ed.), *Problem-based Learning in Education for the Professions*. Kensington, Australia: Higher Education Research and Development Society of Australasia.

Woods, D. R. (2000). An evidence-based strategy for problem solving. *The Journal of Engineering Education, 89*(4), 443–459. doi:10.1002/j.2168-9830.2000.tb00551.x

Worley, C. G., Hitchin, D. E., & Ross, W. L. (1996). *Integrated strategic change*. Reading, MA: Addison Wesley.

Yaman, E., & Güngör, H. (2013). Damgalama (stigma) ölçeği geçerlik ve güvenirlik çalışması. *Değerler Eğitimi Dergisi, 25*(11), 251–270.

Yan-zhong, L. (2005, December7). Fractal administrative organization: New exploration on administrative organizational pattern.*Workshop on Innovations in Governance and Public Service to achieve a Harmonious Society*.

Yell, M. L., Katsiyannis, A., Collins, J. C., & Losinski, M. (2012). Exit exams, high-stakes testing, and students with disabilities: A persistent challenge. *Intervention in School and Clinic, 48*(1), 60–64. doi:10.1177/1053451212449740

Yılmaz Fındık, L. (2015). Postgraduate students' perceptions of leadership in politics: a metaphor analysis sample of Hacettepe university, department of educational sciences. In Ş.Ş. Erçetin & S. Banerjee (Eds.), Chaos, Complexity and Leadership 2013. Springer. DOI doi:10.1007/978-3-319-09710-7_47

Yorks, L., & Kas, E. (2002). Learning from the inquiries: Lessons for using collaborative inquiry as an adult learning strategy. *New Directions for Adult and Continuing Education, 94*(94), 93–104. doi:10.1002/ace.63

Yuan, F., & Woodman, R. W. (2010). Innovative behavior in the workplace: The role of performance and image outcome expectations. *Academy of Management Journal, 53*(2), 323–342. doi:10.5465/AMJ.2010.49388995

Yusif, M. M. (2001). *Education and the war against corruption in Nigeria: some basic issues on the role of primary education*. Academic Press.

Zerbe, W. J., Härtel, C. E., & Ashkanasy, N. M. (Eds.). (2010). *Emotions and organizational dynamism* (Vol. 6). Emerald Group Publishing. doi:10.1108/S1746-9791(2010)6

Zorlu, M & Çalım, S. İ. (2012). İşyerinde damgalama ve ayrımcılık faktörü olarak HIV/AIDS. *Çalışma ve Toplum Dergisi, 4*, 165–188.

About the Contributors

Şefika Şule Erçetin graduated from the Faculty of Economics, Administrative and Social Sciences Department of Business Administration at Anadolu University in 1984. In 1987 she has started to work as research assistant at the Division of Educational Administration, Supervision, Planning, and Economics in Hacettepe University and completed the doctorate there. In 1994 she was invited by The University of Texas at Austin, where she worked as visiting scholar for six months. Since 2003 she has been working as professor. In 2009-2012 she directed many international and national projects about "organizational intelligence, municipalities, vocational education," as project coordinator. She is the president, board member and founder member of the International Science Association (ISCASS). Along with ISCASS, Prof. Dr. Şefika Şule Erçetin organized three international symposiums and now is calling for the new international symposium "4rd International Symposium on Chaos, Complexity and Leadership (ICCLS 2016). She is editor in chief of different national and international journals She has many published articles both national and international journals. She has written seven national books, twenty-four international book chapters and edited seven international books, latter of them published by SPRINGER and IGI Global. She is one of rare academicians in Turkey who interests in "chaos, complexity, quantum leadership" since 1990. She has directed many theses and she gives seminars about these topics. Other main research fields of her are comparative education policies, research methods organizational intelligence and stupidity and wisdom, leadership, migration, fuzzy logic, children at risk, women and peace studies, social media.

* * *

Yusuf Abubakar was born and bred in northern Nigeria, precisely a Sokoto State indigene. His highest qualification is Degree of Master of Educational Management and Administration. Areas of interest are; Human resource management, Leadership and Educational supervision. He contributed in other publications, (Journal). Presently working with ministry of tertiary Education, Mina Niger State Nigeria.

Ali Balcı is currently head of the Department of Educational Administration and Policy at Faculty of Educational Sciences, Ankara University. He received his PhD in Educational Administration and Supervision from Ankara University. He is the author of the books " Effective Schools", "Organization Development", "Organizational Socialization", "Research in Social Sciences", "Educational Administration Terms Glossary", "Organizational Metaphors" (as editor) and "Comparative Education Systems" (as editor). He has more than 40 articles in his own field which has been published in national or international journals. Until now, 40 doctoral dissertations and 38 master theses were completed under his supervision.

Sa'adu I. Bashar was a Master's degree holder in the field of Educational Management and Administration obtained from the Islamic University in Uganda. He is currently a Lecturer in the department of Education Foundations, Sokoto State University, Sokoto, Nigeria. He taught Educational Management and Leadership courses at the undergraduate level and covered areas in relation to organization culture, organizational theories, organizational climate, formal and informal organizations and lots more. He published so many articles and chapters in many international journals, books and international conference proceedings. Currently, he is done with the research on the management of students' discipline in government secondary schools in Isa, Sokoto State, Nigeria.

Ssali Muhammadi Bisaso started as a Teaching Assistant in the faculty of education, department of curriculum and instruction at the Islamic University in Uganda (IUIU) in 2005. He graduated with a Master of Education in Curriculum and Instruction at the International Islamic University, Gombak, Malaysia. He has supervised teacher trainees at school practice and research projects of undergraduate students alongside teaching pedagogy and management laden courses. He also worked as an education consultant with the African Education Trust (AET) and the European Union (EU) in Somali Land and Puntland in 2011. Currently he is undertaking a PhD in Educational Management, Supervision, Planning and Economics at Hacettepe University in Turkey.

Shakira Bodio is an educator in the department of Literature, Language and Linguistics at the faculty of arts and social sciences, Islamic University in Uganda. She has handled teaching and supervision of students especially in the distance learning program and supervised teacher trainees on school practice as well as conducting Micro-teaching sessions for preparation of student teacher-trainees ahead of school practice. She has also contributed immensely to the African children book series by writing pieces of child stories that can be incorporated into language teaching.

Abubakar Mansur Danjeka is an indigene of Bodinga Local Government of Sokoto State, Nigeria. I am currently a lecturer in the Department of Economics, Federal College of Education Zaria, Kaduna State, Nigeria. My Highest qualification is Decree of Master of Educational Management and Administration. My area of specialization is Human Resource Management in Education. The author was married with one wife and two children.

Anıl Kadir Eranıl was born in Nevşehir in 1988 and has been working in the Ministry of National Education since 2010 as a primary school teacher. The bachelor's and master's degree is from Gaziosmanpaşa University and interest is in Leadership, Educational Leadership, Organizational management, Chaos and Complexity.

Barış Eriçok is a lecturer at Nevşehir Hacı Bektaş Veli University, Cappadocia, Turkey. He studied emotional intelligence and managerial competencies of school leaders his master's degree. For his PhD degree, he is currently studying Educational Administration, Supervision, Planning and Economics at Hacettepe University.

Tuncer Fidan is currently working as an internal auditor at Mehmet Akif Ersoy University. He is a doctoral student at the Institute of Educational Sciences, Ankara University. His research interests include organizational socialization, organizational legitimacy, strategic management and positive organizational scholarship.

Leyla Yılmaz Fındık has been working at Hacettepe University as an education and Training planner since 2013. Prior to this job, she worked as a teacher of English at Ministry of National Education between 2005 and 2013. She received B.A. in linguistics at Faculty of Letters from Hacettepe University in 2005; M.A. in 2012 and PhD in 2015 in Educational Administration Supervision Planning and Economics in the Department of Educational Sciences from Hacettepe University. Her primary research interests are in the field of educational administration, leadership, educational planning and higher education and she has several papers on educational administration and leadership. She has worked on different projects related to Bologna Process, ECTS and Students Affairs.

Halime Güngör was born on April 17, 1979 in Bursa in Turkey. She was graduated from Boğaziçi University Faculty of Education Guidance and Psychological Counseling Department in 2003. Then she received her master's degree in Department of Educational Administration and Supervision from Sakarya University, Turkey in May 2013. Her thesis was entitled, "Relationship between Stigma Tendecies of School Principals' and Teachers' and Organizational Culture". She has started to attend Educational Administration, Supervision, Planning and Economics PhD Program since September 2013, in Hacettepe University, Turkey. She is a total quality management leader, school counselor and also managed lots local projects on education.

Kenan Ili graduated from Atatürk University, Department of English Language and Literature in 1996. He has been teaching English since then in State Schools. He completed his master's degree in 2010 at Division of Educational Supervision, Planning and Economics at Hacettepe University. His thesis was entitled, "Impact of Social Media Environments and Tools on Student's Behaviours". He has been at the same division and university as a PhD candidate. He has some academic articles in international journals. His interests are Chaos and Complexity, Educational Leadership, Social Media.

Muhamadi Kaweesi is a full time lecturer and Head of Department of Educational Management and Administration at the International Islamic University in Uganda (IIUIU). In his world of academia, he covers a range of topics in the fields of Educational Leadership and Management, Human Resources Management and development, Project Planning and Management, Organizational Culture in Higher Education, Economics of Education, Organizational Theory and behavior, and Politics and Education. Together with Badru Musisi and Muhammad Bisaso, he developed contextualized study materials for distance learning students in Somalia in different disciplines such as: Principles of Educational Administration and Management, Human Resources Management in Education, Educational Planning, Organizational Behavior in Education, Records Management in Education, Project Planning and Management in Education, Curriculum Management, and Gender in Educational Management. He also operates as an Educational Consultant. His area of research bias is Higher Education.

Mehmet Menteşe graduated from Anadolu University, Faculty of Education in 2002. He is now working as an English teacher at a primary school. He received his master degree at the Hacettepe University Division of Educational Administration, Supervision, Planning and Economics in 2013. He is interested in management, leadership and chaos theory. He continues his PhD at present.

Farooq Miiro is Administrative Officer and instructor in department of educational management and administration, Faculty of Education, Islamic university in Uganda. He joined university from administrative Assistant from 2007-2012, later he was promoted to his current position to date. During this time, he has been a faculty Examinations coordinator for quite some years, through his time several developments have been made for better storage of students' records. He is also involved in community outreaches and sensitization, he has given several public lectures, counseling and guidance at all levels. He did his master's in education management and published a book on Headteacher transfer and the management of secondary schools in Uganda. He is currently pursuing PhD at the International Islamic University Malaysia in same filed of educational leadership and administration. He is interested in engaging in research and publication vigorously in the same field, concentrating more on school leadership and governance, management of higher education, human resource management, child upbringing and school management, working mothers and job performance.

Mukhtar Salihu Nawait is a lecturer in the faculty of Education, Sokoto State University, Sokoto, Nigeria. He graduated from the Dan Fodio University, Sokoto (UDUS) with second class upper in Education (Geography) and later obtained his master degree of Education with specialization in counseling Psychology from the Islamic University in Uganda (IUIU). He taught in Secondary School, lectured in Shehu Shagari College of Education, Sokoto and currently with Sokoto State University. He has written a number of articles in journals and conferences. He is happily married with children.

Mehmet Özbaş is currently Assoc. Prof. Dr. at Erzincan University in Turkey. He holds a BA in Educational Administration and Supervision from Hacettepe University, and MA and PhD in Educational Administration, Supervision and Economy from Hacettepe University. His main research interests are supervision of intra-class activities in elementary education schools, parent-teacher relationships, the reasons for student absenteeism, teaching process depending upon multiple intelligence theory, disadvantageous students and their educational needs, teaching leadership, supervision in educational system, equality of opportunity and possibility in education, social justice in education, the variables that affect the schooling process of female students, poor families and their educational needs, evaluation of the performance of school managers in terms of the conformity with the educational administration scientific area professional acceptance criteria, teacher training. mehmet özbaş studies particularly over educational administration quality, opportunity equality in education and educational rights for the throwaways.

Fathimath Saeed is a lecturer at Villa College, Republic of Maldives. In her courses she focuses on curriculum studies, leadership, management of resources and resource centres, thinking skills and problem solving strategies. In addition, she is also a PhD candidate at Gazi University; Ankara, Turkey. Her department is Educational Management and Supervision. Her PhD research is a comparative study on Education systems of Maldives and Turkey. Moreover, she has also served Maldives education system as a teacher, school principal and Superintendent of Education.

Zayyanu Sambo was a Master's degree holder in the field of Educational Management and Administration. He is currently a Lecturer in the Department of Education Foundations, Shehu Shagari College of Education, Sokoto, Nigeria. He taught Educational Administration and planning course at the NCE and undergraduate levels. He published so many articles and chapters in many international journals, books and international conference proceedings.

Index

O

P

Q

R

S

T

U

W

www.ingramcontent.com/pod-product-compliance
Ingram Content Group UK Ltd.
Pitfield, Milton Keynes, MK11 3LW, UK
UKHW050946290926
15470UKWH00005B/1679